DEVELOPING MATH C   S
in
PRE-KINDERGARTEN

Kathy Richardson

Editing, Design and Layout: Lucinda O'Neill

Illustrations: Linda Starr

Cover Illustration: Peter Zafris

This book is published by Math Perspectives.

Math Perspectives Teacher Development Center
P.O. Box 29418
Bellingham, WA 98228-9418
360-715-2782
www.mathperspectives.com

This book is printed on recycled paper.

Distributed by Didax, Inc.

Didax
395 Main Street
Rowley, MA 01969-1207
800-458-0024
www.didax.com

B C D E F 13 12 11 10 09

ISBN 978-0-9724238-9-2

Table of Contents

Setting 2: Circle Time, p. 64

Setting 3: Teacher-Directed Small-Group Time, p. 99

Developing Math Concepts in Pre-kindergarten

Setting 4: Individual Learning Time, *p. 137*

Setting 5: Learning with Others Throughout the Day, *p. 163*

Appendix A: Planning Guide for Pre-Kindergarten Math Time, *p. 171*

Appendix B: Blackline Masters, *p. 187*

Introduction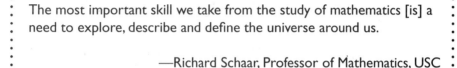

The mathematician and the preschooler have much in common. They approach the study of mathematics in much the same way. Like mathematicians, young children are intrigued by the mathematics in their world. Consider the intensity and determination children bring to their investigations of the mathematics that surrounds them. This comes not from a need to complete an assignment or because it will be useful to them. It comes from a need to know.

Young children are naturally interested in numbers and spontaneously ask: *How many are there? How many do we need? Do we have enough? Who has the most? Are there any extras?*

They are interested in geometry and explore to find out: *How are these shapes alike? How are they different? Which ones fit together? Which ones leave spaces? What can we build with these? What other shapes can we make using these shapes?*

They want to measure and compare, and they wonder: *Which is bigger? More? Heavier? Longer? Shorter? How can we find out?*

They experience the connections between math and music when exploring rhythm and patterns, and between math and art when working with symmetry and design.

The mathematics that engages the mathematician and the preschooler is the study of patterns and relationships, order and predictability. This is not mathematics the way most of us experienced it in school. Yet it is the mathematics that will serve our children most as they move through school, and, in time, that will await them as they enter the workforce. The mathematician and the preschooler need and want to experience the joy and excitement of the world of mathematics.

Mathematics in Pre-kindergarten: What does it look like?

The learning of mathematics is an active endeavor. Children need to be involved in investigating, comparing, wondering, and checking to see what happens. They need to think about what they are experiencing, to notice what happens, and to begin to talk about what they notice. They need to listen and think about what other people have to say. They need to begin to make connections, to see relationships between mathematical ideas. They need to see how experiences can be recorded: with blocks, with pictures, and sometimes with mathematical symbols. Important mathematical ideas will naturally arise through children's play.

> Children's play and interests are the sources of their first mathematical experiences. These experiences become mathematical as the children represent and reflect on them...The most powerful mathematics for a preschooler is usually not acquired while sitting down in a group lesson but is brought forth by the teacher from the child's own self-directed, intrinsically motivated activity.
>
> —Douglas H. Clements, *Mathematics for Young Children*

Children's mathematics includes:

Number experiences ...

that ask them to find out: *How many? How many leaves did you pick up? How many pockets do you have? Can you put three horses in the corral? Can you put four fish in the ocean? Can you give me four of those?*

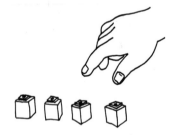

Acting out counting songs helps bring meaning to the counting sequence they are learning. Children learn to count beyond what they already know by counting along with the teacher.

"How many children are here today?" the teacher asks, and models writing numerals when the children tell how many.

Some children will label what they counted with a numeral and show how many horses they put in the corral, and a few will even try to write the numeral on their own.

Spatial experiences ...

that allow them to see how shapes fit together through experiences with puzzles and blocks and boxes. Children have many experiences working with blocks, and they build structures that begin to look more and more organized and stay balanced a bit longer.

They hear their teacher using language that describes where objects are located:

"Here it is—under the table."
"Let's look behind the cubbies."
"We put the calendar over the bookshelf."

And they begin to use spatial language themselves, like

"The baby is in the baby bed. She is under the covers."
"The cookies are inside our play oven."

Pattern experiences ...

that focus on figuring out what comes next. The teacher begins a rhythmic pattern, and the children join in. "Clap, clap, floor, floor, clap, clap, floor, floor."

The teacher models patterning when she makes a long Unifix train with her students and asks: "Red, red, yellow, red, red, yellow. What color do we think comes next?"

The teacher notices and points out to her students the patterns on their clothing: "Look, I see a pattern on your shirt. Red, green, white, red, green, white."

Children are given opportunities to copy repeating patterns using a variety of materials— for example, block, button, block, button. If they have pattern cards available to copy, some will continue the pattern beyond the card.

Children experience patterns through music and become aware of rhythm and repetition when they march around the room, tap their rhythm sticks, or beat on their drums. They experience patterns through stories when teacher reads predictable books, and they delight in guessing what comes next.

Measurement experiences ...

that focus on making comparisons between objects. "This pumpkin is bigger than this one." "This one is really heavy." Children find things to compare in all their daily activities: "My blocks go higher than yours." "I made a long, long train."

They begin to go beyond describing relationships only as *big* and *little*, and they hear the teacher use terms like *heavy* and *light*, *long* and *short*, *wide* and *narrow*, *thick* and *thin*, *tall* and *short*, *large* and *small*, and *more* and *less*.

Some children will line up blocks to look like stair steps or put yarn lengths in order from the shortest to the longest. It is not always an easy task, and they line them up carefully to see which is longer.

Data collection experiences ...

that focus on sorting and counting as a means of finding answers to everyday questions like:

"How many boys are in school today?"

"How many children had a turn on the slide?"

"Here are all the leaves we collected on our walk. Let's put the green ones here and the brown ones here. Did we find more brown ones or more green ones?"

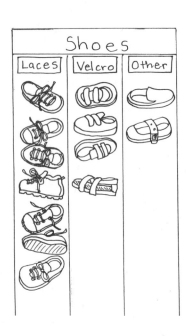

How Children Learn Math Concepts

The National Council of Teachers of Mathematics included prekindergarten standards for the first time in its 2000 *Principals and Standards for School Mathematics*. The NCTM document outlines the mathematics that children should learn as they progress through school, identifying both content and process standards.

The content standards are organized into five areas: Number and Operations, Algebra, Geometry, Measurement, and Data Analysis and Probability. The process standards are Problem Solving, Reasoning and Proof, Communication, Connections, and Representation.

The mathematics presented in *Principles and Standards for School Mathematics* provides a broad view of what mathematics is and can be for young children. The NCTM standards show that mathematics can provide children with ways to understand and appreciate the world around them and enrich their experiences.

The following sections highlight what children need to know in each of the math strands and how they learn the mathematics that are and will be important to them.

Learning Number Concepts

In pre-kindergarten through grade 2, all students should

- **Count with understanding and recognize "how many" in sets of objects;**
- **Connect number words and symbols to the quantities they represent.**

—NCTM *Principles and Standards*

The complexity of the number concepts young children must encounter and sort out is not always immediately observable to the adults in their world. We sometimes get clues from children, yet do not realize what these clues are telling us. A child brings a package of gum to school, hoping to share it with the entire class. Another child puzzles his teacher by bringing one pencil when the teacher asked for four. When asked to show why he thought he was supposed to bring that pencil, the child counts, "One, two, three, four. See, this is four," and hands the pencil "named" four to the teacher.

We watch another child counting his toy cars and see him pointing to some more than once and skipping others. "This time I have nine," he says. Another child, with great enthusiasm, shares, "My dog had a hundred puppies!" These same children may know how to count by rote to 20 or know the numerals to 10. But as we can tell by these examples, there is more to understanding what number is all about than rote counting and numeral recognition.

As useful as these skills may be, we want much more for our children than that. The most important underlying number concept for children to begin to develop in pre-kindergarten is a sense of quantity. This is not something easily attained and certainly not mastered in pre-kindergarten, but it is the search for the sense of number that is worthy of the child's time.

So, we will not be content even if children can count to 20 or recognize numerals. We want them to be able to use counting when they need to find out how many. We want them to begin to become aware of when an answer is reasonable or not reasonable. We want them to become more consistent and accurate when they count. We want them to begin to see relationships between numbers. And, most of all, we want children to believe that numbers make sense and to be confident about their own abilities to deal with them.

In order to provide the kind of environment that allows young children to develop a sense of number confidently in their own time and way, we must understand something about how this happens for them. Young children do not see the world of number in the same way adults do. They do not yet trust that quantities remain the same when they appear to be different. They are not necessarily aware that they must keep track of what they have already counted so some items won't be counted twice.

It is a natural part of their development for children to believe that quantities are unstable and unpredictable. It is not helpful to them to be told that their incomplete understanding is wrong—that they should know, for example, that the quantity doesn't change just because it is rearranged. If they are told that they should know what they are not yet able to know, they lose faith in their ability to make sense of things and instead begin early to play the game "Don't think—just figure out what the teacher wants you to say."

True understanding of number must be developed in the child's mind as a result of his or her own experiences and reflections on those experiences. As teachers, we must resist the temptation to tell and explain. Simply provide opportunities and ask questions that get children to look and to think. Showing or telling children will not make them understand. They must figure things out for themselves.

Number sense can only develop if children are given many and varied experiences of finding out "How many?" and are allowed to come to an understanding about number in their own time and way. We provide opportunities to look, to find out, to notice, to ponder, to wonder, and to think about numbers.

Number is an idea that a child must understand about the objects being counted that goes beyond their physical properties. The following basic concepts are those that young children are confronted with as they seek to make sense of numbers: quantification, one-to-one correspondence, conservation of number, relationships between numbers, and symbolization.

Quantification: When children first learn to talk, they are very interested in labeling things. They learn "bottle" and "ball" and "baby." But numbers are not labels for particular objects. To understand number, children need to understand that the number word they say includes all the objects previously counted. We see a lack of understanding of this idea when children respond to the request "Give me four" by picking up the one object they were pointing at when they said "four" while counting.

One-to-one correspondence: When children are first learning to count, they have a general notion about pointing while they say the counting sequence "One, two, three, four, five…" and they imitate the counting behavior they have observed. At first, they are neither very precise nor too concerned about making sure each object is counted once and only once. So we see children saying the counting sequence faster or slower than the speed at which they are pointing to the objects being counted. Children do not really understand what counting is all about until they are able to consistently count each object (physically or mentally) once and only once.

Conservation of number: What is often most surprising to adults when they attempt to understand children's development of number concepts is seeing how children are misled by their perceptions. When children have not yet developed an understanding of conservation of number, they believe that the number changes when the appearance of the group of objects changes. They believe that they have more objects if they are spread out and thus fill up more space. They think they have more if their sandwich is cut into four pieces instead of two pieces. They are not sure what happens to the quantity of objects if some are hidden. They may believe there are more chairs than pencils because the pencils are smaller.

Relationships between numbers: Children do not automatically see the relationships between numbers. They don't see that four is contained in and is a part of six. Even when children know how to count, they do not necessarily know what happens when one is added to a set of objects. They often need to count to see. It is especially difficult for them to know, without counting, how many are in a group of objects if one has been removed. Additionally, they do not know that when sets are reorganized in different ways, the quantity stays the same. So they don't realize, for example, that 4 and 2 is equivalent to 1 and 5 or even 2 and 4.

Symbolization: Children will often deal with symbols as though they were physical realities. They see the symbol "3" literally as a squiggle on paper. They don't see it as a mark representing something else. It is important then to help them associate symbols with the quantities they represent. Students will learn much about representing numbers with numerals through the opportunities that naturally arise when teachers find it necessary to use numerals and to model their use in context. When you write the number of days the children have been in school, you will be using numerals. When you record attendance or keep track of the number of children buying milk, record the numerals in a way that children can see them. The main goal is for the children to see numerals as useful tools for writing down information. Many will be interested and will pick up the names of the numerals even from this indirect exposure. Others will be becoming familiar with the peculiar shapes and forms that make up the numerals. And others will benefit from the small group work where symbols are used to write down "How many?"

Learning Geometry Concepts

In pre-kindergarten through grade 2, all students should

- Recognize, name, build, draw, compare, and sort two- and three-dimensional shapes;

- Describe attributes and parts of two- and three-dimensional shapes;

- Investigate and predict the results of putting together and taking apart two- and three-dimensional shapes;

- Describe, name, and interpret relative positions in space and apply ideas about relative position;

- Recognize and create shapes that have symmetry;

- Recognize geometric shapes and structures in the environment and specify their location.

—NCTM *Principles and Standards*

The study of geometry helps us look more closely at the physical world in which we live. When we observe children who are intrigued and challenged when building with blocks, constructing towers, or creating symmetrical designs, we are seeing them engaged in the study of geometry.

> In the study of shape, our goals are not so very different from those of the ancient Greek philosophers: to discover similarities and differences among objects, to analyze the components of form, and to recognize shapes in different representations. Classification, analysis, and representation are our three principal tools.
>
> —Marjorie Senechal, *On the Shoulders of Giants,* p. 140

In their study, they are finding out how two-dimensional and three-dimensional forms of all kinds fill up space, how they stack, and how they fit together. As children work with blocks of various kinds, as they create and copy designs and structures, as they examine and analyze boxes and containers, they become more and more discriminating.

Rather than simply learning the names of the basic shapes, they learn to recognize the attributes of shapes, to notice how shapes are alike and different. They learn to identify and sort by such attributes as the number of sides, the number of corners, the number of faces, the relationship of lengths of sides, whether the figure has straight lines or curves, whether it rolls or doesn't roll, and whether it has symmetry or not. They learn what things change and which stay the same when direction, position, or size is changed.

As they explore their world, children discover forms and structures that occur over and over again. Some of these will be simple forms like circles or rectangles, but they also will become aware of more complex forms, such as pyramids or spirals.

Often the focus of geometry in the early years is on learning the names of shapes. However, it is important to recognize that the name of a geometric figure not only names but defines that figure. The name carries with it the particular attributes of the shape. For example, when we know what a square is, we also know it is a four-sided figure that has four right angles with all four sides the same length. However, when children learn the name of a shape without understanding what attributes define that shape, they can end up with misconceptions.

For example, children who think of △ as a triangle often think that ▽ is not a triangle because it is upside down. Some will consider ◣ as half a triangle because it looks like half of the triangle that is familiar to them. If children think a rectangle is the shape that looks like a door, they won't realize that ▭ is also a rectangle. And they won't recognize that a square is a rectangle, too.

Sometimes adults are impressed when children use mathematical language, but they are not always aware that children do not fully understand the words they are using. For example, when children learn to call the yellow pattern block a hexagon, it appears that they are becoming more proficient at using mathematical language. However, children are not usually learning what about that block is or is not relevant to its being called a hexagon but are just naming that particular piece. If the yellow pattern block ⬡ is the only hexagon children experience, they won't realize that ⎍ is also a hexagon and they may mistakenly think ⬭ is a hexagon because it looks like a hexagon.

Before children can make sense of geometric language and apply it appropriately, they must be able to distinguish and label those attributes that define the shape. In the process of learning the language, they will be more apt to focus on the attributes if they are asked to describe the attributes in their own words, using whatever language makes sense to them.

The first step in geometric language development should be learning to see, to notice, to discriminate. The next step should be determining which shapes with similar attributes go together and why. Children should learn the formal labels only when they are ready to apply them to many different versions of a particular shape. Learning the language prematurely can only cause confusion and misconceptions and keep children from looking closely at important attributes. Simply naming geometric shapes is not what is important. Rather, it is important that children look carefully at the properties of various shapes and learn to distinguish among them.

It is very important that children have experiences exploring blocks and the other math materials. It is through exploration that children discover and explore mathematical relationships. Block play is more important than some might realize. According to the research, pre-kindergarten children who are able to build complex structures with blocks have a better chance of mathematical success in middle and high school, even taking into account students' IQ levels, social class, and gender (Wolfgang, Stannard, and Jones, 2001).

Learning Sorting and Classifying Concepts

In pre-kindergarten through grade 2, all students should

- **Sort, classify, and order objects by size, number, and other properties.**

—NCTM *Principles and Standards*

Sorting and classifying is an important intellectual process that we go through when we seek to organize our world. Classifying requires us to create mental structures based on complex relationships and interrelationships. Children begin the process of identifying classes of things early in life as they learn to identify chairs in all their forms and to distinguish them from the various kinds of couches, or to tell a horse from a cow or a dog from a cat. I remember the first time I realized how complex this could be when my own young daughter pointed out a fluffy little animal and said, "Look, there's a kitty dog."

Even though the process of organizing the world through sorting and classifying begins very early in the child's life, it is a process that takes many years and proceeds through many stages. Children's beginning work toward an understanding of classes and classification includes sorting and forming collections using real objects. In order to sort, children must be able to recognize and identify attributes of the materials being sorted. Next, they determine which attributes are alike (which things go together when sorting) and which are different. Their ability to do this grows with time and experience. Through handling materials, they will come to notice finer and more detailed distinctions. They will learn to attend to one particular attribute. They will learn to ignore the differences and focus on what makes a group of objects alike. As children develop their awareness, their language will also develop as the need to be more precise increases.

Children generally first identify the color of objects and later notice shape, size, and other characteristics. Their first intentional sorting behavior is often one of finding pairs that are just alike. Another step in sorting is finding all the objects in a collection that have a particular attribute, while leaving those that do not. For example, they may pull out all the red ones from a group. All the rest that are not red are left in the original pile.

It isn't until around age six that most children will be able to sort all the objects in a set according to a particular property. For example, they will be able to find all the round ones and then go on to find all the square ones and then continue to sort all the rest of the shapes in their collection. Before this there is a transition stage, where children start with one attribute and move to another in the middle of sorting. For example, a child might start sorting all the red ones and then start making piles of circles. Their focus is more on finding things that are alike than on determining a classification that would encompass all the objects they are working with.

One reason that sorting is so challenging for children is that until they reach a certain stage, they cannot think of more than one attribute at a time. For example, it is difficult for them to consider an object as both red and round. If they are trying to sort objects into particular sets and find something that could go in two different places, they often just ignore it.

Learning Pattern Concepts

In pre-kindergarten through grade 2, all students should

- **Recognize, describe, and extend patterns such as sequences of sounds and shapes or simple numeric patterns and translate from one representation to another;**

- **Analyze how both repeating and growing patterns are generated.**

—NCTM *Principles and Standards*

Much of our lives is spent in an active search to make sense of things—to organize and sort things out. When we are able to get a sense of the basic order of things, we are able to predict—to count on things happening—and thus to become more secure and confident. We learn that morning always follows night, that second grade comes after first, that a flash of lightning will be followed by a roll of thunder.

Seeing patterns in the way things work is an incredibly powerful learning tool that most of us have developed intuitively to some degree. If we add s to words to make them plural, if we know a dog we have never seen before is a dog and not a cat, if we know that red lights always turn green and that green lights turn yellow, we are using our sense of pattern.

Mathematicians say that mathematics is the study of pattern. Pattern is the basis on which our number system was created. For many of us, mathematics has not been about patterns but rather a series of rules and steps to follow so we could get answers to teachers' questions. We did not try to look for the underlying order or sense of things in mathematics. When we have not discovered the pattern in numbers, learning mathematics has been much more difficult than it needs to be, and we missed much of the beauty that is in mathematics.

We can give our students a sense of the beauty and order that is mathematics. We can give them the confidence that comes when things are predictable. The activities in this book are designed to give children such opportunities.

Children move through several stages as they develop an understanding of pattern. At first, many do not know what is being referred to when the term *pattern* is used. They often have misconceptions and incomplete understandings as they seek to figure out what a pattern is.

Once children begin to recognize the underlying order and predictability in the patterns they experience, they begin to create their own very simple patterns. The first patterns children create are frequently simple AB (alternating) patterns, so it might seem logical to present AB patterns first and make sure children understand these before moving on to more complex patterns. However, when we try to teach children in a step-by-step fashion, we sometimes limit their view. They are likely to end up with a misconception of what a pattern is, falsely concluding that all patterns are AB patterns. This confusion may not show up until children try to apply their notions about pattern to a new situation.

If we want children to fully understand what a pattern is, we must surround them with a variety of patterns of many forms and in many different situations. Children learn to make connections and to see relationships when they are immersed in experiences—when they see an idea portrayed in many different ways. It is important that we present the variety of patterns to children from the very beginning and not try to oversimplify the idea for them.

Learning Measurement Concepts

In pre-kindergarten through grade 2, all students should

- **Recognize the attributes of length, volume, weight, area, and time.**

—NCTM *Principles and Standards*

Young children's interest in measurement is primarily focused on finding out who or what is the biggest. One of the important concepts they need to learn is that "big" can mean many different things depending on the property being considered. A jar may be bigger if we are talking about height but smaller if we are talking about how much it holds. We might consider one child to be bigger than another because he weighs more, or another child might be considered bigger because he is taller.

When we measure, we need to decide what we want to know and then choose the tool that will help us find out. Do we want to know how long something is, how tall it is, how much space it takes up, how much it holds, or how heavy it is? Learning which property is being measured can be challenging to the young child. Young children who are just beginning to work with measurement often mix up the language, revealing the difficulty of the ideas they are exploring. For example, one child lay down next to a long Unifix train he had made and declared, "This is how big I weigh."

One of the underlying concepts that influences the young child's understanding of measurement is the idea of conservation. Conservation is the recognition that a quantity or amount stays the same even though it has been rearranged in some way and appears to be different. This seems obvious to an adult but is not at all obvious to a young learner. For example, at a certain stage of development, a child will say that he has "more cracker" because he has broken it into pieces.

A child once told me, "I can finish my whole lunch. My mom used to cut my sandwich into four pieces, but that made too much. Now she cuts my sandwich in two pieces, and I can eat it all!" Children reach these kinds of conclusions in part because they generally can attend to only one attribute at a time.

For example, if two sticks are lined up as in the following illustration (left), children will see they are the same length.

However, when one is moved (right), children may focus on only one of the end points and thus think the stick in front is now longer.

When children pour water from one short, wide jar into a tall, narrow jar, many think the taller jar holds more. When a ball of clay is rolled into a "snake," children think the quantity has changed and is either more or less depending on the attribute on which they are focused.

If children are truly at this stage of thinking, they cannot be shown the right answer. What they perceive to be true is more powerful in their thinking than what someone else might try to get them to understand. When we ask children to work with measurement concepts, we must take into consideration the influence their understanding of conservation has on their understanding of measurement and look at their responses in light of this stage of thinking.

As children develop an understanding of measurement, they move through predictable stages of development. Their first understanding of measurement requires them to compare things directly to see what is "bigger" and what is "littler." Eventually, they will see that, in some situations, it is difficult to line things up. They will then be encouraged to find tools to help them measure. The first tool they experience might be a piece of string they use to measure the length of an object. The idea of units of measure will begin with finding out how many scoops of rice it takes to fill a jar or how many cubes an object weighs. These kinds of experiences along with questions and conversations will prepare children to work with standard units and tools when they are presented in the months or years ahead.

Learning Data Collection Concepts

In pre-kindergarten through grade 2, all students should

- **Represent data using concrete objects, pictures, and graphs.**

—NCTM *Principles and Standards*

One of the ways we use mathematics to understand our world is through the collection of data and the organization of that data into graphs and charts. When information is organized into graphs, the visual arrangement of the information reveals much about a situation and makes that information more easily accessible to us.

We can help children become familiar with graphs and their purposes if we provide them with opportunities to collect and organize data to answer questions. The questions asked should be those that have some interest to young children and that deal with their immediate world. Mathematics and science, as well as mathematics and social studies, are integrated naturally when we gather and collect data.

Teaching Mathematics in the Pre-K Classroom

Teaching math in the pre-kindergarten classroom depends on creating a learning environment that supports the learning of mathematics. This includes not only the physical layout of the room but also the settings in which children work, the kinds of tasks in which they are engaged, and the ways in which teachers work with their students. All of these factors must be considered as we work to create a supportive learning environment.

The learning environment we are aiming for is one that
- engages children's thinking,
- provokes questions,
- stimulates a search for meaning,
- encourages children to look for connections and relationships, and
- helps children make sense of the mathematics in which they are engaged.

The Teacher's Role

(Teachers) ... provide an environment that provokes puzzlement, stimulates curiosity, encourages a sense of wonder, and prompts questions and investigations. Within such an environment, educators need to interact with young learners in ways that model thinking and problem-solving approaches and challenge children's existing ideas about how things work.

—Rosemary Althouse, *Investigating Mathematics with Young Children*, p. 4.

The mathematics that children learn depends in large part on what experiences their teachers provide and the interactions the teacher has with each child. It is the teacher who identifies and highlights the mathematics the children are experiencing by helping them become aware of the mathematics that surrounds them.

The teacher helps the children to become aware of the mathematics they are experiencing in several ways,

sometimes by noticing ("I see that ..."),

sometimes by asking ("What did you find out? What do you think will happen if ..."),

sometimes by modeling ("Look what happens when I ...").

Teachers create an environment that encourages thoughtfulness and curiosity. When the teacher not only provides the learning opportunities but also asks children to think about and describe what they are learning, many ordinary events can become important in learning mathematics. The teacher needs to capitalize on every opportunity. One way to do this is by not doing things that children can do for themselves. For example, when materials need to be handed out, children can be asked to bring just enough for the others in their group. Even clean-up time can be a learning experience. Putting the building blocks away requires children to sort and match and make things fit in the space allotted.

The way teachers interact with their students is key to establishing an environment that encourages thinking. When you present a problem to children, be sure you are prepared to accept their ideas even if they come up with some that are totally unexpected. It is important that we ask questions that don't have preconceived "right" answers. When we know exactly what we want our students to say, the game becomes "guess what the teacher wants."

Rather than focusing on asking children to use precise mathematical language, it is more important for them to be thinking and using language that is natural to them. When they are wrong, instead of simply correcting them, ask questions such as "How can we find out?" or "Who has a different idea?"

In addition to informal situations that lend themselves to mathematical learning, the teacher also has the job of providing more formal learning opportunities in which the learning goals and time to practice are specifically planned.

> If curriculum depth and coherence are important, then unplanned experiences with mathematics clearly are not enough. Effective mathematics programs also include intentionally organized learning experiences that build children's understanding over time. Thus, early childhood educators need to plan for children's in-depth involvement with mathematical ideas.
>
> —NAEYC and NCTM Position Paper

Providing Appropriate Instruction

Today, teachers are being encouraged to bring more mathematics to their pre-kindergarten classroom than in the past. With this greater emphasis on mathematics, it is important that curriculum practices and assessment techniques stay consistent with what is known about how young children learn and the way their development affects their learning.

If young children are to benefit from more mathematical experiences, it is essential that the focus be on a search for meaning and understanding. Mathematics programs must enhance and maximize the child's learning rather than require children to work with math in ways that give the appearance of high expectations but that, in reality, result in inappropriate practices that interfere with their growth.

As stated in the *Principles and Standards for School Mathematics*, "Learning with understanding is essential to enable students to solve the new kinds of problems they will inevitably face in the future" (NCTM, 2000, p. 21). It is too easy to get children to say and do what will please an adult even though it has no meaning for them.

It is imperative that we focus on the mathematics that will best prepare young children for the work they will be asked to do in the future. Sometimes what teachers think will help children most when they enter kindergarten is not in fact what they really need to know. I observed this for myself when I visited a group of kindergarten children who had been placed in a class designed to provide them with extra help in mathematics. Most of them had attended a preschool for at-risk children before coming to kindergarten, and their teacher was eager to provide them with meaningful and motivating experiences.

> Children as young as two may be able to count to 10 or even to 20, but if they do not link their counting to quantification, it is no different from memorizing their ABCs or a list of names like Bob, Joe, Sara.
>
> —Eugene Geist, "Children Are Born Mathematicians," p. 14

The teacher had asked the children to bring teddy bears to school so they could measure them. The children snapped Unifix Cubes together into trains that were about as long as their bears. But the teacher couldn't get them to tell how many cubes long their bears were. The children were struggling with this, and the young teacher was baffled. She knew the children had learned to count in preschool. Most of them were able to count to 20, and many of them recognized numerals to 10. What she hadn't found out yet was that very few of them could actually count more than three or four objects. They simply could not count the trains they had made to measure their bears.

What these children had learned in preschool had value, but it was not what they needed most in order to make the expected progress in kindergarten. They needed to learn to count to find out "how many." They needed to see how counting was used and to develop meaning for the numbers they worked with.

Children who have been identified as at risk, who leave their pre-kindergarten experience appearing to know what is necessary but without the underlying ideas to build on, will continue to fall behind if these needs are not met.

We do not need to give children experiences that "look like" what they will have to know later. Children do not need to learn about rulers to learn about measurement. In order to understand measurement, children first need to be aware of what can be measured. They need to line things up, to cover spaces with blocks, and to pour sand or water from one container to another. Learning the names of shapes is not the best preparation for understanding the geometry ideas they will eventually need to know. If children are going to understand geometric principles, they first must put together blocks to make new shapes and to recognize the difference between a triangle and a rectangle.

Rather than focusing on how "high" a child can count, the emphasis in prekindergarten needs to be on helping children understand quantities. Children need to count objects to find out "How many?" They need to experience that 4 is 1 more than 3 and that 2 is less than 6. They need to ponder the fact that 5 objects can be arranged in different ways: as 3 and 2 or 4 and 1, and also as 2 and 2 and 1, and still be 5.

Language Development

The development of mathematical language should come from the myriad of experiences a young child has throughout the day. It is important to recognize that the development of language begins with the need to communicate, not with the need to learn new words. The central role of the teacher is to encourage children to look closely and to notice things. Children develop language skills when the teacher is interested in what they are seeing and doing and asks questions that create the need for communication. It is from this search for ways to share experiences that language grows. The focus should be on the natural language of the child before more formal or standardized language is expected.

Teachers need to be aware of the specific language they want children to learn and to model the use of this language as opportunities present themselves. Children will then learn to use this language when they have something they want to express.

Position words that are basic to language in pre-kindergarten should be modeled in settings where children are required to look at relationships. Teachers need to use words and phrases like *above, below, before, after, high, low, in front of, in back of, inside, outside, on top of,* and *under* to highlight relationships and to give these words meaning.

There will also be opportunities to model language when children are working with shapes and when measuring and comparing. *Round, curved,* and *straight* are words that children should hear in addition to *circle, square,* and *triangle.* And they can learn to associate what is being measured with the words that describe particular attributes, like *height, weight, tall, thin, wide,* and *narrow.*

Rich experiences, careful observation and opportunities to communicate will naturally lead to rich language.

The Young Child's Perspective: What do children understand?

Undeveloped ideas and misconceptions are a normal part of the child's evolving understanding. One student of children's learning tells of his experience when checking for understanding of the idea of conservation. The teacher whose class he was observing had worked with her group of young students to help them understand that just because a jar looks taller, it does not mean it can hold more water. She demonstrated how a tall, skinny jar holds less than a short, fat jar. When the researcher visited the classroom, the children were able to tell him that the short, fat jar held more than the tall, skinny jar. However, when he offered them lemonade and asked them to choose between the lemonade in tall, skinny glasses and the lemonade in short, fat glasses, all of the children chose a tall, skinny glass. They had learned to respond correctly to an adult's questions, but their behavior revealed they were not totally convinced.

We can't prevent misconceptions like this or ensure that children are learning what they need to learn by teaching them to say words or perform procedures they don't understand. If we want children to make sense of mathematics, we must provide a variety of experiences that ask them to think about what they are doing and help them focus on the critical elements of concepts. It is through encountering an idea in different settings and in many different ways over time that generalizations begin to form. If we insist that children must always have correct responses for concepts they are not ready to understand, they must resort to rote memory of these correct responses because they will not be able to make sense of the situations by themselves.

We often make assumptions that children are thinking what we are thinking when they perform correctly. For example, I remember a time when I had my pre-kindergarten class work with dot cards. I would show the children cards with the same dot arrangements as dice, and they learned to recognize these arrangements. One day, I asked them to use counters and build what they saw on the card. To my amazement, they did not use the correct number of counters. Instead they made an X shape to match the shape of the five dots, and they made a "squarish" shape to match the arrangement of the nine dots. I thought I was teaching them quantity, but they were focused on what the card looked like.

I learned from this experience that I must always interact with children in ways that ask them to show me what they know rather than assuming they are thinking what I am thinking. Children can say the right answer and not know what they are saying, as with the dot cards. They can also give the "wrong" answer and still be pondering an important idea.

When giving a child a task or a question, we do not need to know ahead of time whether it will be out of reach or not. If we are willing to learn from children's honest responses, we will be able to present ideas to them in all their complexity rather than oversimplifying them in order to ensure "success."

Another experience comes to mind. A group of children were asked to figure out which jar held the most rice. In the set were three mustard jars that were the same shape but different sizes. One of the children was not convinced that the jars should all hold the same amount of rice. Even after pouring the rice from one jar to another, he said, "I see it. But I don't believe it!" This response indicates that the child is still searching and trying to make sense. He is not content with accepting something just because "teacher said."

When we attempt to teach children our way of thinking or our way of getting answers before they can understand, we only interfere with their sense-making process. They stop looking for their own meaning and instead look to the teacher to see if they are right or wrong. As long as teachers think their job is to make sure children "do it right," their students will be limited in their ability to understand and make sense of concepts.

The focus should be on wondering, finding out, and coming up with ideas. Over time, the child who is thinking and noticing will refine and increase his understanding as he seeks to make sense of his world.

Differentiating Instruction: Meeting the Range of Needs

> With the enormous variability in young children's development, neither education policy makers nor teachers should set a fixed timeline for reaching each specific learning objective.
>
> —NAEYC and NCTM Position Paper

It is critically important that we acknowledge and support each child's level of thinking and understanding. As I worked with children over the years, I had to redefine what it meant to be a "good" teacher. I learned that being a good teacher is not about getting all your students to perform at a particular level at a particular time. Being a good teacher is about knowing what your students already know and what they are still grappling with. It is valuing where each child is on his own personal journey rather than comparing him to someone else.

Standards and goals give teachers an idea of what we must be working toward with our children. However, one child may be a long way from achieving a particular standard, while another child may have reached that standard long ago. Our job as teachers is to challenge all our students, no matter if they are just figuring out how to choose two jelly beans or if they are figuring out how to share 12 jelly beans with two people. If we value each child's learning, instead of being disappointed that she did not perform at the same level as another child, we can be excited by the process she is engaged in.

We need to be aware of the accomplishments of each child in the class. When watching children at block play, we can appreciate the stages of learning they are moving through as we see one child making an elaborate castle, while another child lines up two blocks. We see that one child tries to draw a picture of the plate of cookies and write the numeral telling how many, while another child counts the cookies with the teacher.

David Elkind (1999) reminds us that, as we seek to determine what is possible for children, "the only way to understand how children learn a concept is to observe them in the process of acquiring it." The learning experiences included in this book inherently allow for a range of needs. Not all children are expected to get the same thing from the same experiences. The size of the numbers, the complexity of the patterns, the nature of the language used can all be varied in response to what the teacher observes as the children work and explore.

Children do not need to be grouped by ability in order to perform the individual activities. Individual needs can be met as children at many different levels of understanding work side by side. It is important that the task cards include a variety of levels of complexity, as the tasks are made harder or easier depending on the complexity of the patterns being worked with. Some children will choose simple patterns to work with, while others will choose more complicated patterns.

When children work on the same activity but at different levels, they come to accept that everyone has his or her own "work to do" and do not judge their work as "ahead" or "behind" another child's work. There is no need for children who are still working with simple patterns to feel inadequate. All tasks can be valued and accepted as worthy endeavors as long as the appropriate effort is being made. Providing the opportunity for children to work on the same activities at different levels does much to promote their acceptance of a variety of tasks when working together.

When children work independently, you can learn much about their thinking and level of development by observing them at work. You can interact with them individually, providing support and challenge as needed. You can also pull out small groups of children who may need extra support or an additional challenge.

Setting Up the Classroom

The Physical Layout of the Room

Children need big spaces in which to work more than they need little areas that divide up the room. Plan for a large rug area for Circle Time. This can also be a place where the children can build with building blocks and explore other math materials. You also need to have assigned places where the math materials are stored so that the children will be able to help set up and put things away.

Gathering and Preparing the Materials

It is important that children use a variety of materials when working with mathematical ideas. The following are suggestions for materials that are available and can be used in many ways. It is not necessary to have exactly these materials. Feel free to use whatever materials you have or can easily get.

Basic Materials:

- Large building blocks
- Small building blocks such as geoblocks, Discovery Blocks
- Plain wooden cubes
- Colored tiles
- Pattern blocks
- Attribute blocks
- Geoboards and geobands
- Unifix® or other interlocking cubes
- Polygon shapes

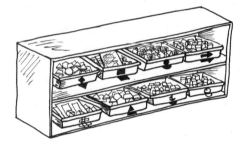

Things to Gather:

- Various jars and other clear containers
- Various objects to weigh
- Small mirrors
- Collection of lids (from peanut butter jars, mayonnaise jars, margarine tubs, and so forth)
- Collection of small boxes of various sizes and shapes, with and without lids (Examples: small jewelry boxes, shoe boxes, cereal boxes, cracker boxes)

Collectibles:

- Rocks, seeds, beans, pompoms, shells, acorns, bottle caps, buttons

Organizers:

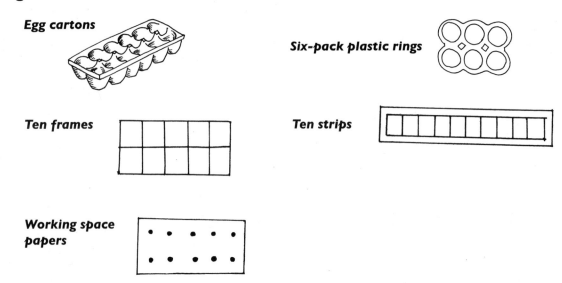

Egg cartons

Six-pack plastic rings

Ten frames

Ten strips

Working space papers

Things to Make: "Let's Pretend" Materials

Children have great imaginations, and it is important to give them many opportunities to exercise them. A whole world can be opened up to a child with the slightest suggestion. "Let's Pretend" materials are a very simple set of materials that can be used in a variety of ways. Pieces of paper or simple constructions serve as "environments," and Unifix Cubes or plain wooden cubes, or collectibles such as rocks, pompoms, or lima beans, can stand in as things that could be found in that particular environment. The children will be able to think of many of these possibilities themselves.

The following "Let's Pretend" materials can be used for counting, acting out finger plays, making up story problems, and symbol matching.

Blue paper:	*Green paper:*
Environment: sky, water, lake, pond, ocean, and swimming pool	*Environment:* meadow, playground, park, baseball field, football field
Unifix Cubes or plain wooden cubes can stand for: airplanes, kites, balloons, birds or fish, boats, children, sharks, frogs	*Unifix Cubes or plain wooden cubes can stand for:* children, ball players, ladybugs, mice, bees

Sand paper: **Environment:** beach, sandbox **Unifix Cubes or plain wooden cubes can stand for:** shells, crabs, clams, toys, children (or you can use real shells)	**Black paper:** **Environment:** night, outer space, cave **Unifix Cubes or plain wooden cubes can stand for:** witches, ghosts, pumpkins, rockets, bats
Waxed paper: **Environment:** ice, North Pole, frozen pond **Unifix Cubes or plain wooden cubes can stand for:** skaters, penguins, polar bears	**White paper:** **Environment:** snow **Unifix Cubes or plain wooden cubes can stand for:** snowmen, snowballs, children, sleds

Things to Make: Number Sets

The number-set materials consist of objects permanently arranged in sets from 1 to 10. While it is important to have a variety of these sets of materials, it is not necessary to have all the variations. Each set should include at least three or four of each of the numbers from 1 to 10. Some possible materials for making these sets are:

Buttons glued down on cards

Foam stickers glued on cards

Plastic paper clips glued down on cards

Bread tags glued down on cards

Toothpicks glued down on cards

Beans glued into portion cups

Foam beads strung on stretchy nylon and tied to make bracelets

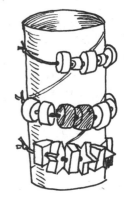

Pompoms glued into portions cups

Holes punched in long, narrow cards

Beads strung on yarn

Objects glued on sticks

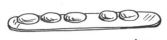

Beads on pipe cleaners glued to sticks

Stickers glued onto a ten frame

 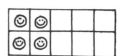

Beads cut from long strings

All of these number sets can be used for counting, matching, sorting, comparing, and combining. Some of the materials lend themselves better than others to certain of the activities. For example, beads are especially effective for one-to-one counting because children can move them. The button cards are more of a challenge for sorting because they are made in random arrangements and require children to count carefully.

Most of the number-set activities are written to be done with any or all of the number sets. When a specific type of number-set material is needed, it will be specified in the particular lessons.

References

Althouse, R. (1994). *Investigating mathematics with young children*. New York: Teachers College Press.

Clements, D. H. (2003). *Mathematics for young children*. Presentation at the National Head Start Development Institute, Washington, D.C.

Elkind, D. (1999). Educating young children in math, science, and technology. In American Association for the Advancement of Science, *Dialogue on early childhood science, mathematics, and technology education*. Washington, DC: AAAS. ED 427 877.

Geist, E. (2001, July). Children are born mathematicians: Promoting the construction of early mathematical concepts in children under five. *Young Children, 56* (4), 12–19.

Holt, J. C. (1983). *How Children Learn*. New York: Perseus Publishing.

NAEYC and NCTM (2002). *Early childhood mathematics: Promoting good beginnings: A joint position statement of the National Association for the Education of Young Children (NAEYC) and the National Council for Teachers of Mathematics (NCTM)*.

NCTM. (2000) *Principles and Standards for School Mathematics*.

Reifel, S. (1984, November). Block construction: Childen's developmental landmarks in representation of space. *Young Children*, 61–64.

Schaar, R. (2003). *NCSM Fall Newsletter*, p. 26

Senechal, M. (1990). Shapes. In L. A. Steen (Ed.) *On the shoulders of giants* (pp. 139–182). Washington, DC: National Academy Press.

Wolfgang, C., Stannard, L., & Jones, I. (2001). Block play performance among preschoolers as a predictor of later school achievement in mathermatics. *Journal of Research in Childhood Education, 15* (2).

About the Activities

The activities and experiences described in this book are organized according to the settings in which they are presented. This is to make it easier for teachers to find the type of activities they want to present.

This section describes the general learning goals for each of the math strands. A list of activities for each strand is included for teachers who wish to access the activities by particular math concepts.

The Number Concept Activities

With these activities, children will have the opportunity to:

*   Practice rote counting to 10 and beyond
*   Learn one-to-one counting to 10
*   Develop a sense of numbers and number relationships
*   Recognize numerals 0 to 10

These counting activities are designed to give children opportunities to make sense of number. The activities challenge them to think about numbers and to notice what happens when they count, but they are not expected to memorize things they do not understand.

Most of the time, the focus should be on the development of counting skills and a sense of the numbers being worked with. Sometimes children will be asked to compare two sets to see if one is more, less, or the same as another set. Children will also have opportunities to learn to associate groups of objects with the numerals that describe them.

The Number Concept Activities are:

Circle Time	How Many Days in School?, p. 69
	Counting Poems and Finger Plays, p. 74
	Feel the Beat, p. 76
	Counting Jars, p. 78

The Geometry Activities

With these activities, children will have the opportunity to:

- Develop spatial sense
- Explore two- and three-dimensional shapes and structures
- Describe the attributes of shapes and structures
- Take shapes apart and put them together to create new shapes
- Discover and analyze shapes in the real world
- Change and compare shapes
- Examine positions and relationships between positions
- Develop useful language

The activities presented here will help children develop geometry concepts and spatial awareness over time. These particular tasks have been chosen to give children the various kinds of experiences they need to build a base for further geometry work in later years. The activities can be performed at various levels of difficulty and should be available for children to work with many times. The level of complexity will change as children continue to revisit them.

The Geometry Activities are:

Exploration Time	Exploring Building Blocks, p. 48
	Exploring Boxes, p. 51
	Exploring Small Blocks, p. 52
	Exploring Counting Materials, p. 53
	Exploring with Mirrors, p. 54
	Exploring Geoboards, p. 58
Circle Time	Geometry Walks, p. 91
Small-Group Time	Copying Designs on a Geoboard, p. 130
	Grab Bag Geometry, p. 131
Individual Learning Time	Copy Cat, p. 155
	Matching Lids and Boxes, p. 157
	Recording Designs and Creations, p. 158
	Cut-and-Paste Geometry, p. 159
	Pattern Block Puzzles, p 161

The Sorting Activities

With these activities, children will have the opportunity to:

- Identify attributes
- Decide how things are alike and how they are different
- Decide which things go together and why
- Sort and re-sort according to different attributes
- Label sets according to the way they were sorted

Children will need to sort a variety of things in the normal course of the day. In fact, clean-up time can be one of their most educational experiences. This time will be made more valuable if you discuss where things go and why, eliciting first the children's ideas about different places that the materials might go.

Most of the sorting activities will occur in small groups, where you will be able to focus the children's attention on attributes and the language that can be used to describe those attributes. Notice that many of these activities could be included in the geometry section as well, since the focus is on sorting by geometric attributes. This is just one example of the connections that exist among all the areas of mathematics.

The Sorting Activities are:

Exploration Time	Exploring Collections, p. 56
Circle Time	Alike and Different, p. 80
	Descriptions, p. 82
Small-Group Time	Alike and Different, p. 80
	Descriptions, p. 82
	Sorting Attribute Blocks, p. 124
	Sorting Collections with the Teacher, p. 125
	Sorting Shapes, p. 126
	Sorting Blocks, p. 127
	Sorting Boxes, p. 128
Individual Learning Time	Sorting Collections Independently, p. 154

The Pattern Activities

With these activities, children will have the opportunity to:

- Recognize repeating patterns
- Copy and extend patterns
- Create patterns
- Discover patterns in the real world
- Be introduced to growing patterns

Young children need to experience patterns first as motion, color, design, and arrangement. Later, from these experiences, comes the discovery of the patterns that exist in number. The complexity of the patterns experienced in the Pattern Activities can be varied according to the needs of individual children. The level of complexity that the children will be able to copy and extend will change over time as they learn to work at a higher level.

The following are examples of the types of patterns young children can recognize, analyze, and extend:

Rhythmic Patterns

(Touch) *Nose, nose, knees, nose, nose, knees …*

Clap, clap, slap, slap …

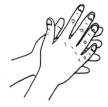 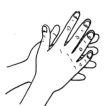  

Color Patterns

Green, green, yellow: green, green, yellow …

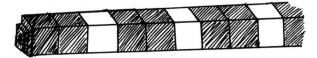

Shape Patterns Art

Triangle, triangle, triangle, circle…
Triangle, square, square…

Patterns in Kind

Button, tile, tile …

Patterns in Design Art

Pine tree, pine tree, sun, sun …

Position or Arrangement Art

Up, down, sideways …
Over, under, over, under …

Attributes Art

Big, big, little …

Bumpy, smooth, bumpy, smooth …

Children should experience patterns in an ongoing way in various classroom settings. They can be introduced to patterns when the teacher presents a variety of patterns during Circle Time and can work on copying and extending patterns during the time spent at math stations.

Encourage your students to make long patterns, because it is through repetition that children begin to get a true sense that patterns go on and on. Individual needs can be met even when children at different levels work side by side, because the task cards include a variety of levels of complexity. Some children will choose simple patterns to work with, while others will choose more complicated ones.

The Pattern Activities are:

Circle Time	What Day Is It Today?, p. 66
	Rhythmic Patterns, p. 70
	People Patterns, p. 84
	Patterns in the Environment, p. 86
Small-Group Time	Patterns in the Environment, p. 86
Individual Learning Time	Pattern Station Task Cards, p. 151
Throughout the Day	I See a Pattern!, p. 168

The Measuring Activities

With these activities, children will have the opportunity to:

- Explore measuring lengths
- Compare two objects to see which is longer and which is shorter
- Compare various items to a particular item to see which are longer or shorter
- Explore measuring capacity
- Compare two objects to see which holds more and which holds less
- Compare various containers to a particular container to see which holds more or less
- Explore measuring weights
- Compare two objects to see which is heavier and which is lighter
- Compare various items to a particular item to see which are heavier or lighter

The activities presented here provide children with opportunities to explore measuring length, capacity, and weight. Some of the measuring tasks can be done by the whole class working at the same time. All of the children can search for things that measure the same as their foot, or use string to find things that measure the same as a particular classroom object. The Measuring Jars Station and the Weighing Station, however, can be used by only a few children at a time. If children are to have enough time at these stations, the stations need to be available on an ongoing basis throughout the year.

In addition to the measuring activities presented during planned lessons, young children need many informal experiences if they are to develop an understanding of measurement. They need opportunities to explore their world and to continually ask the questions: "How can we find out what's the same?" "What's longer, heavier, bigger, holds more?" "What's shorter, lighter, smaller, holds less?"

The value in these experiences comes from the active involvement of children and their interactions with the ideas they are exploring. Learning does not come from experiencing any particular task once but through many experiences with the same activity. It is through this process that children begin to notice more and to see relationships previously unseen.

Measuring Length

Children's earliest measuring experiences include standing back to back to see who is taller and lining up toys to see which is longer. Working with length, then, is a natural starting point for learning to measure. Children begin their work by finding objects in the classroom that are the same length as one of their own body measurements. They then compare objects to find items that are longer or shorter than this measure. For example, they may search for things that are longer or shorter than their arm, foot, or thumb.

The Measuring Activities related to measuring length are:

Circle Time / Small-Group Time	What's the Same as Me?, p. 93
	Is It Longer or Shorter?, p. 95
	Measuring with Strings, p. 97
Throughout the Day	Let's Measure It!, p. 169

Measuring Capacity

Children will work with measuring capacity using the Measuring Jars Station. When they measure capacity, they will be surprised by many of their results. They will see that tall jars often hold less than short jars. Two jars that look very different may hold the same amount. They will explore these sometimes puzzling relationships first by pouring the contents of one jar into another. Children need time to explore the Measuring Jars Station independently. They will work with the teacher when the focus is on developing the appropriate vocabulary and on recording what they learned.

The Measuring Activities related to measuring capacity are:

Exploration Time	Exploring at the Measuring Jars Station, p. 59
Small-Group Time	Measuring Containers, p. 133
Individual Learning Time	Exploring at the Measuring Jars Station, p. 59
	Measuring Containers, p. 133

Measuring Weight

The children will begin their work with weight by weighing two objects to see which is heavier and which is lighter. Next, they compare several objects to see if they are heavier, lighter, or the same as one particular object.

The measuring activities related to measuring weight are:

Exploration Time	Exploring at the Weighing Station, p. 61

Small-Group Time	Comparing Objects Using Scales, p. 134
	Is It Heavier or Lighter?, p. 135
Individual Learning Time	Exploring at the Weighing Station, p. 61
	Comparing Objects Using Scales, p. 134
	Is It Heavier or Lighter?, p. 135

The Data Collection Activities

With these activities, children will have the opportunity to:

- Gather information
- Organize information in order to see relationships
- Describe the relationships that graphs reveal

Young children's experiences with data collection and graphing should involve them in gathering and organizing real information. When numbers refer to real things, they inherently carry meaning and help children see the usefulness of numbers.

First experiences collecting data should involve getting answers to everyday questions. For example, the teacher asks, "How many children chose an orange for lunch?" or "How many books did we check out at the library?" Children count to find out the answer and aren't yet asked to do any comparing.

Then, with the teacher's help, children can gather and organize objects, pictures, or name cards into graphs so that they can see the relationships that exist. To see these relationships, they will need to put these items into one-to-one correspondence so that the graph itself displays which group has more and which has less.

We want children to eventually understand that they don't always have to count items to see which is more and which less. They can accomplish this by lining things up. We also want them to see that it matters how we line things up if we are going to see the relationships easily.

Discuss this with the children in an ongoing way as various situations arise. For example, if you are doing a graph of the fruit the children brought for fruit salad, have them consider how they could line up the different pieces of fruit to make it easier to tell if there are more bananas or more apples. Remember, what is obvious to you may not be obvious to them. They may think that there are more bananas than apples because the bananas make a longer line:

Using the activities in this book, children can experience different types of graphs. One type of graph is a Real Graph. Comparing objects themselves is a very real and visual display of the data and the relationships among them.

Another type of graph is a Picture or Name Graph, in which information about an individual child can be gathered. A symbolic graph is difficult for young children to relate to. An *x* on a graph is abstract and, to young children, seems disconnected and unrelated to them. Children need experiences that allow them to see themselves in the graphs. One way to do this is to have children actually place their name or picture on the graph. They can then see the collected information but also know they are a part of that collection.

Even though most of the graphs that young children experience will come out of daily activities and interests that come up in the classroom, there are some data that can be collected over time. Keeping track of the weather, for example, in a Weather Graph will give children a sense of collecting data over time using information that is very real to them.

All data collection activities are whole-group, teacher-directed activities. The Data Collection Activities are:

| Circle Time | Weather Graph, p. 73 |
| | Real Graphs / Name Graphs, p. 88 |

Settings for Learning

4

Mathematics can be presented in five different classroom settings:

1 During the exploration of materials,

2 During teacher-directed whole-group circle time,

3 During teacher-direct small-group learning time,

4 Through individual learning experiences available at different learning stations in the classroom,

5 Through a variety of natural opportunities that occur throughout the day.

Each of these settings provides a different but important way for children to learn mathematics.

Setting 1: Exploration Time

Much of young children's learning occurs when they work with math materials using their own ideas. To make sure children have the time they need to work with these materials, a particular time should be set aside each day during which they are allowed to discover the potential of the materials and to follow their own ideas.

> Children need what we rarely give them in school—time to build up in their minds without hurry, without pressure, ... a mental model of the territory before they start trying to talk about it. We teachers like to think we can transplant our own mental models into the minds of children by means of explanations. It can't be done.
>
> —John Holt, *How Children Learn*, p. 222

What may look like play time is actually critical to the work children will do at a later time when they use those same materials in the more structured tasks you will have them do.

Setting 2: Circle Time

Circle Time is an opportunity to present activities to all the children at one time. This helps them begin to feel like they are part of a community of learners and allows you to make sure that all the children have particular experiences.

Setting 3: Teacher-Directed Small-Group Time

Small-group time is the setting in which you provide learning experiences that are not as effective when done with the whole class but that the children are not yet able to do independently. You can also use this time to introduce and reteach those activities you want the children to be able to do independently.

Setting 4: Individual Learning Time

In this setting, the children choose from among tasks that are set up at different learning stations in the classroom and are available for several weeks at a time. The children will benefit from doing these tasks over and over again.

Setting 5: Learning with Others Throughout the Day

Opportunities will arise naturally throughout the day that lend themselves to counting, observing relationships, and solving problems. It is important to take advantage of these opportunities whenever possible.

For more information on structuring the day to include all five settings for learning, see the Planning Guide in the Appendix (page 173).

Setting 1 : Exploration Time

Before children can work on the particular tasks the teacher has in mind for them, they must be given the opportunity to explore materials in their own way. Focusing on the properties and attributes of materials is important work for the young child. Exploring math materials is how they learn and discover relationships and "how things work." They build mental models of shapes and dimensions and become familiar with the critical characteristics of the various materials they are working with. They need to touch and move and stack and fit things together and will often come up with more insights than we might have assigned.

There is much for children to learn, so they need time to use their own ideas and follow their own curiosity all year long. Their work will become more complex and more organized over time as their awareness of the potential of the materials deepens.

The benefits of Exploration Time are many:

Children explore and discover mathematical ideas.

When we watch children working with math materials using their own ideas, we see examples of their creativity and problem-solving abilities. But beyond that, we need to look for evidence of the mathematics learning that is naturally occurring. As children become familiar with the attributes and possibilities of the materials, they will naturally:

- Create pictures and designs
- Sort in a variety of ways
- Measure and compare
- Build increasingly complex structures
- Encounter problems of balance and symmetry
- Create and extend patterns
- Count and compare quantities
- Use words and phrases like *longer, shorter, not as much, more, just the same*

Children learn to work with others.

When children work with math materials in their own ways, they not only learn about the mathematics inherent in the materials but also become familiar with the culture and expectations of the classroom. During Exploration Time they will be learning the importance of:

- The appropriate use of materials
- Working with others: sharing materials and space
- Choosing tasks and staying with them
- Setting up and putting away

Children learn appropriate behavior for working with classroom materials.

It is very natural for children to work creatively with materials. They know what to do even without direction from the teacher. However, the tone you set will have a big influence on the attitude with which they come to the task. Emphasize the importance of working hard. Since I don't want to imply that this is playtime or a time for children to do whatever they want, I often use the phrase, "working hard using your own ideas" rather than saying, "You can do what you want with the materials."

There are a few rules that the children need to understand if this time is to go smoothly. I share the following basic rules with the children.

- **We do not throw things in the classroom.** (Our classroom needs to be a safe place.)

- **We share the materials.** (We don't need to hoard. We take only what we need as we need it.)

- **We clean up before moving to a new station.** (This leaves the station ready for other children so they can have a turn using the materials.)

- **We mess up our own work only.** (When people create something, they usually want to take it apart themselves. When someone else messes up our work, it feels like they are wrecking it.)

The rules for using classroom materials need to make sense to the children. They need to see that the rules help everyone get along better with each other.

Exploration Time Activities:

Exploring Building Blocks (Geometry), p. 48

Exploring Boxes (Geometry), p. 51

Exploring Small Blocks (Geometry), p. 52

Exploring Counting Materials (Geometry), p. 53

Exploring with Mirrors (Geometry), p. 54

Exploring Collections (Sorting), p. 56

Exploring Geoboards (Geometry), p. 58

Exploring at the Measuring Jars Station (Measurement), p. 59

Exploring at the Weighing Station (Measurement), p. 61

Exploring Building Blocks

Materials:

• Large building blocks

Lesson Focus:

• To explore, build, and create using the blocks

The Task:

Introduce the blocks to the children. Ask: "What can you do with these blocks? What do you notice about them?"

It is through actual work with blocks that children become aware of the structure of the various kinds of blocks and their geometric attributes. The children should be encouraged to compare, sort, and arrange objects according to shape, color, and size, and to develop notions of similarities and differences. The children will create increasingly complex designs and structures when allowed to work with the materials over time.

Interactions: Deepening the Understanding

Your way of interacting with and your expectations of your students will play a big role in the level at which they work. At times you will want to interact with them to help them focus on the mathematics inherent in what they are doing.

Sometimes you will make observations about what they have built, offering comments such as "I notice here you made a wall with a small rectangle and then you made another wall that is twice as big" or "I see you made it look the same on both sides."

There may be times when children will need a suggestion or will want you to join them in a task if they are unable to get going on their own. Children will feel free to decide on their own whether to follow your suggestions if you state them in a manner such as

"I wonder if ..." or

"Do you think it would work to ... ?" or

"Do you have another idea?"

Or you could say: ***"I wonder if you could make a road with that kind of block."***

To further focus the children on their tasks, ask them questions and make observations such as:

"Can you tell me what you are working on?"

"Tell me about what you have made."

"What did you do when you were making it?"

"I noticed you used this (type of) block before. What other kinds of blocks did you use?"

"I noticed you used long blocks for this part of your wall and short blocks for this part of your wall."

"How many blocks did you need to build this side? What about this side?"

Observations: Watching the Growth

How complex are the structures your students build? Over time, have they developed an increasing sense of orderliness, balance, and symmetry in their creations? Are they able to tell about their structures? Do they describe them in terms of what they look like ("I made a castle" or "I made a tower"), or do they describe them in terms of how they built them?

You can learn much about your students' ability to work with geometric ideas simply by watching them at work. When children first begin to play with blocks, they make simple rows or stacks of blocks. Young children learn to build more complex structures when they have opportunities to work with blocks and develop their skills.

The illustration on the next page, from the journal of the National Association for the Education of Young Children, shows the natural stages that children go through with experience. (Reifel, 1984)

Children go through all the stages of block positioning as shown. In any pre-kindergarten classroom, there will be children who are just discovering what they can do with blocks. Others will be able to build much more complex structures. One of the milestones achieved by most four-year-olds is the ability to build and represent enclosures.

a. stack, for *on* (vertical)

e. enclosure (flat)

b. row, for *by* (horizontal)

f. enclosure (arches)

c. stack and row combination

g. enclosure (combination)

d. pile three dimensions with no interior space

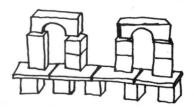

h. combinations of many forms

Exploring Boxes

Materials:

- Boxes of various sizes and shapes: small jewelry boxes, shoe boxes, cereal boxes, and cracker boxes, and so on (with and without lids)

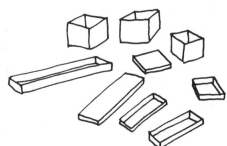

Lesson Focus:

- To become familiar with the geometric properties of boxes through independent exploration.

The Task:

Introduce the boxes to the children. Ask:

"What can we do with these boxes?"

"What can we discover?"

Interactions: Deepening the Understanding

It will be natural for the children to focus first on what came in the boxes, saying things like "This is for jewelry," "I know this is for shoes," or "This must be a candy box." Some will comment on the physical attributes as well, saying things like "This one can stand up" or "This is a pointed box."

The children will explore the boxes in various ways. Some children will sort the boxes, some will build with them, and some will fit them inside each other. Give the children the time they need to explore the boxes in their own way. Let them tell you what they notice at this stage, whether it has anything to do with geometry or not.

Observations: Watching the Growth

What kinds of comments do the children make?

Do they notice more about the attributes of the boxes after several experiences with them?

Exploring Small Blocks

Materials:

- Pattern blocks
- Geoblocks
- Tiles
- Wooden cubes
- Tangrams
- Discovery Blocks

Lesson Focus:

- To become familiar with the geometric properties of various types of blocks by exploring and building with them

- To explore how the different types of blocks stack and fit together

The Task:

Introduce the blocks to the children. Ask:

"What can you do with these blocks? What do you notice?"

"Can you tell me about what you make? What did you make? What did you have to do in order to make it?"

Interactions: Deepening the Understanding

Ask:

"Which shapes fit together with no spaces between?"

"Can you build a floor with one kind of block? With lots of different kinds of blocks?"

"Can you build a road? Is it going to go straight or turn?"

"What shapes can you make with other shapes?"

Observations: Watching the Growth

How complex are the children's structures and designs?

Over time, have they developed an increasing sense of orderliness, balance, and symmetry in their creations? Are they able to tell about their structures?

Do they describe them in terms of what they look like ("I made a castle" or "I made a tower"), or do they describe them in terms of how they built them ("I put a little block on each corner of the big block")?

Exploring Counting Materials

Materials:

- Any materials you plan to use for counting, such as Unifix cubes, color tiles, wooden cubes, toothpicks, number sets

Lesson Focus:

- To explore and become familiar with the attributes of the materials.

The Task:

Introduce the counting materials to the children. Ask:

"What can you do with these? What do you notice about them?"

It is important that children have opportunities to work with all the materials they will be using later for various counting tasks. They will be more ready and willing to do what you have in mind if they are already familiar with the materials and have experienced working with them in their own ways.

Interactions: Deepening the Understanding

If they need encouragement to get started, say:

"I wonder what you could make with these?"
"I wonder if you could make a tower? A train? A picture?"

Ask them about their work:

"Can you tell me what you are working on?"
"Tell me about what you made."
"What did you notice about these?"

Share what you noticed: ***"I see that you ..."***

Observations: Watching the Growth

Are the children able to think of things to do? Do they get more ideas over time?

Are they making long structures, tall structures, pictures?

Are they ordering or making patterns?

Exploring with Mirrors

Materials:

- Small mirrors
- Books
- Teacher-made task cards showing letters and shapes

Lesson Focus:

- To explore mirrors and see what happens when books, letters, and shapes are reflected in a mirror

- To become familiar with the ideas of symmetry and reflection

- To develop vocabulary that describes shapes, characteristics, and discoveries using the child's own language

The Task:

Introduce the mirrors to the children. Ask: ***"What do you notice when you work with the mirrors?"***

Developing Math Concepts in Pre-kindergarten

Interactions: Deepening the Understanding

After the children have had opportunities over several days to explore the mirrors in their own way, you can guide their observations while preparing them for ideas of symmetry. Ask:

"When I put the mirror here, does the (book, letter, shape) look the same or different?"

"Look at the T (or M). Can you find a place to put the mirror so you still have a T? An M?"

"Can you make it look different than a T (or M)?"

"What happens when you put the mirror here on the circle? What happens when you put the mirror on a triangle? What are the different shapes you can make?"

"Look at this letter. What happens when I put the mirror here? What do you see?"

Observations: Watching the Growth

What is the level of interest in the mirrors?

What do the children notice and comment on?

Are they simply exploring to see what happens, or do they seem to be deliberate in their attempts to figure something out?

Are they surprised?

Do they want to share their discoveries?

Exploring Collections

Materials:

- Collections of various items, such as buttons, keys, pods, and nuts and bolts

(Collections that contain many duplicates are easier to sort. For more challenging sorting, you may want to include items with fewer duplicates.)

Lesson Focus:

- To become familiar with the attributes of the materials by exploring, describing, arranging, and sorting the items in various ways

The Task:

Introduce the collections to the children and tell them to see what they can find out about the items in the collection.

Interactions: Deepening the Understanding

While they are working, focus the children's attention on the attributes of the various items in the collections:

"Can you tell me about this button? This key (and so on)?"

"I see a button with two holes. Can you see any buttons with two holes?"

Help develop vocabulary:

"I see a button that has a ridge like this. Do you see any buttons with a ridge?"

"I heard someone say 'shiny.' Can you find buttons that are not shiny? Some words for 'not shiny' could be 'dull,' 'rough,' or 'scratched.'"

It might be helpful to make a list of the words the children use to describe objects so that you can review their discoveries and share those descriptive labels with the whole group at a later time.

Observations: Watching the Growth

What do the children do with the collections?

Do they arrange them in a design?

Do they find exact matches?

Do they describe what they see?

Can they describe more than one attribute of an item at a time? (For example, "This is a button, and it is a circle, too.")

Exploring Geoboards

Materials:

- Geoboards
- Geobands

Lesson Focus:

- To create shapes and designs by placing geobands on geoboards

The Task:

Introduce the geoboards and geobands to the children. Model how to put the geobands on the geoboards. Ask:

"What can you make on the geoboard using these geobands?"

"What could you make if you could use only one or two geobands?"

Interactions: Deepening the Understanding

Ask:

"What did you make? Can you copy this on another geoboard? Can you draw it on paper?"

"Can you think of something in the room that has that shape? What is a good name for that shape? Can you make different kinds? What is the same and what is different?"

Observations: Watching the Growth

Do the children simply pile on geobands, or are they intentional in trying to create a design?

Do they stop when they recognize that they have made something, or do they continue to put more and more geobands on the geoboard?

Can the children describe their shapes using their own language? (For example, "It looks like a fat L.")

Exploring at the Measuring Jars Station

Materials:

- 1 or 2 tubs of rice (or other pourable material)
- Jars of various sizes and shapes labeled with colored dots
- Various-sized scoops
- Funnels (can be made by cutting off the top of a two-liter soft drink bottle and turning it upside down)
- Something to catch spilled rice (could be a big box with the sides cut down to 4 or 5 inches, a sheet of oil cloth, or even a baby's swimming pool)

If there is a water table in the classroom, children can compare jars by pouring water from one jar into another.

Lesson Focus:

- To explore what happens when filling jars with rice and pouring from one jar into another

The Task:

Introduce the Measuring Jars Station. Model guidelines for using the jars and rice.

Model questions and experiments that you would like the children to pursue when exploring capacity. For example, hold up two jars and ask: *"Which jar do you think holds more—the jar with the red dot or the jar with the green dot? How can we find out?"*

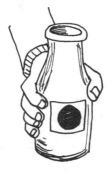

Fill one jar and pour the contents into a second jar. Ask:

"What happened? Can we tell which jar holds more and which jar holds less?"

Explain to the children that they will have opportunities to work independently with the jars at a later time, and establish rules for exploring the jars:

"Hold the jars you are filling over the rice tub so the rice that spills will fall into the tub."

"Don't pour from so high up that the rice bounces all over."

"Clean up the spilled rice carefully when you are finished working at the station."

Interactions: Deepening the Understanding

At first, children will naturally focus on filling the jars. To help children move past this, ask:

"What do you notice about the jars?"

"Which jar do you think holds the most?"

"Can you find two jars that hold the same?"

"Which jar holds less than the others?"

Observations: Watching the Growth

Are the children focused primarily on how the rice feels and the act of pouring from one jar into another?

Do they seem to be paying attention to the relationships between jars, or are they simply intrigued with filling up the jars?

After time and experiences, do they comment on the jars and tell what they have noticed?

After the children know how to work with the rice and jars, provide ongoing experiences for them to explore at the station. Encourage the use of language that describes the discoveries they are making.

Developing Math Concepts in Pre-kindergarten

Exploring at the Weighing Station

Materials:

- Homemade scales (instructions included at the end of this activity)

- Objects to weigh, such as rocks, apples, oranges, potatoes, clay, small cans of food, or balls, and small containers of things such as beans, rice, sand, or popcorn

  (Be sure to include some objects such as Styrofoam or pumice so that children can experience that size doesn't always equal weight.)

Lesson Focus:

- To become familiar with the Weighing Station

- To work independently with the scales

- To explore aspects of "heaviness" and "lightness" by seeing what happens when objects are placed in the scales

The Task:

Demonstrate how to weigh objects using two of the homemade scales.

Model what you would like the children to do when exploring weight. Choose two items to weigh and ask:

"Which do you think is heavier—the can or the rock? Let's find out."

Place the objects in the scales and have the children report what happened. Tell the children they will be able to work with the scales themselves at a later time.

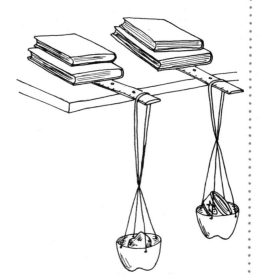

Establish the rule for using the scales in this first lesson, by saying:

"Put only one thing in each scale at a time."

Interactions: Deepening the Understanding

When children first work with scales, they often want to fill them as full as possible to see how far down the scales will go. (This is a parallel to filling all the jars at the Measuring Jars Station.) At this stage, they are experiencing the "heaviness" of objects and are not necessarily focused on comparing them. To help children move past this stage, ask:

"What have you found out about what you weighed so far?"

Observations: Watching the Growth

Are the children aware of differences in weight, or are they just intrigued with the experience of weighing without paying particular attention to what they are weighing?

Do they begin to compare objects intentionally?

Do they make comments on what they have found out?

A Note about Homemade Scales

When children are first learning about weight, they need the type of scales that show how the weight of an object actually pulls down the scale. This helps them more easily see the heaviness of the object. For that reason, we recommend using something like the homemade scales described on the next page. A commercial spring scale has some of the same attributes, but at this stage, it distracts the children from what we want them to observe.

Making the Homemade Scales

To make each scale, you will need:

- One plastic two-liter soft drink bottle or similar container
- Two rubber bands (stretchy enough to weigh light things like bars of soap, a rock, washers, and so on)
- An old ruler
- Four pieces of string, each about 11 inches long

1. With a knife, cut the bottle in half crosswise, and then trim with a scissors along the bottom edge of the label, making a base for the scale that measures about 2.5 inches high. Discard the middle section of the bottle.

   (You can also trim the bottle along the top of the label and set the top portion aside to be used as a rice funnel.)

2. Mark the placement of 4 holes, which will be punched for the strings. Make sure they are spaced evenly and opposite one another.

3. Punch out the holes with a compass point, ice pick, or nail.

4. Make a knot at the end of each string large enough to keep it from going through the hole.

5. Thread each string through a hole with the knot inside the base.

6. Bring all four strings together and tie with a good knot.

7. Attach a thin rubber band around the tied strings, looping the rubber band through itself with the string caught in the middle. This makes it easy to replace if the rubber band breaks or is too strong to show differences in weight.

8. Place the ruler on top of a bookcase, table, or desk, and weigh it down with a heavy object.

9. To use the scale, hang it by looping the rubber band over the ruler.

Setting **2**: Circle Time

There are certain types of activities that are most appropriate for whole-group time. Some will be presented as daily activities and others will be shared experiences in which children can work at their own pace.

Ongoing Activities

Some Circle Time activities are an ongoing part of the day and become part of the daily routine, such as opening activities, rhythmic patterns, practicing the counting sequence, using the Counting Jars activity (p. 78) to estimate and count, and graphing. However, because children need to be involved in doing their own work as much as possible, keep this time short.

Get individual children thinking as much as possible so that they are not simply responding along with the group. For example, all the children can be involved in estimating the number of walnuts it takes to fill a jar. Even though it is the teacher who fills the jar while the children count the walnuts, there is a sense of active participation as they watch the jar being filled and compare what they thought was going to happen with what actually happens.

Be aware of how you work with the whole group, especially if you find that the whole class is responding as one voice. A group response may be appropriate, but usually it indicates a memorized response that individual children may or may not understand. One of the biggest problems with whole-group work is the assumption we often make that all (or even most) of the children understand what they are saying in chorus. It will only be in other settings like independent learning time that you will be able to tell if these words have any meaning for them.

Shared Experiences

These are activities that the whole class can do at the same time if there are enough materials for all the children to work with at one time. For the tasks to be appropriate for the whole group, individual children should be able to do the same task in their own way. They should not be kept together or directed step by step through the tasks. For example, the children can work, each in their own way, on activities such as finding objects that are as long as their foot or making patterns with tiles.

During whole-group activities, it is not always possible to meet the specific needs of particular children. It is extremely important, therefore, to provide other opportunities for each child to be involved with the materials and be personally challenged. Individual children's needs can be met very effectively during small-group, focused learning time and during independent learning time at the different learning stations.

The following activities work well during Circle Time:

Daily Routines:

Learning the Counting Sequence:

Sorting and Patterns:

Data Collection:

Geometry and Measurement:

What Day Is It Today?

Materials:

- Blank calendar for each month
- Calendar symbols for each month (made by the children)

Lesson Focus:

- To become acquainted with the calendar
- To experience a sense of the passing of time
- To learn the days of the week
- To become familiar with the names of the months

The Task:

Have the children create different symbols, such as paper cut-outs of apples, clouds, or geometric shapes. Then, each school day, write the date on the appropriate symbol for that day and add the symbol to the calendar. (Days not in school will be added all at once on the first day back.)

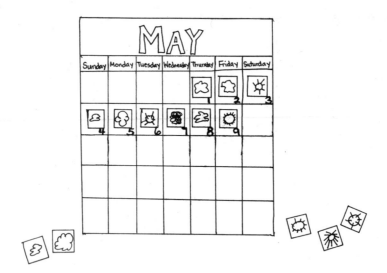

Because the calendar is an ongoing experience, it provides an opportunity for children to see a pattern emerge as the symbol for each date is placed on the calendar. For example, say:

"Look at the pattern we are making on our calendar. Cloud, cloud, sun; cloud, cloud, sun; cloud, cloud, sun … What comes next?"

"Cloud!"

"What number do we need to write? Let's count and see. One, two, three, four, five, six, seven, eight, nine ... What's next?"

"Ten!"

As the months go by and you continue to add to the calendar, your students will become more and more familiar with the number sequence to 30.

Interactions: Deepening the Understanding

Ask questions that invite the children to notice the pattern as it emerges:

"Let's say the pattern together and clap every time we say 'sun.'"

"Let's say the pattern together and see if you can tell me what I am covering up. Cloud, cloud, sun, cloud, cloud, sun. What is under my hand?"

"What do you think the pattern will be tomorrow? And after that, what will come next?"

"What number will we write tomorrow? What number did we write yesterday?"

Observations: Watching the Growth

Are the children able to join in? With ease or difficulty?

Can they predict what part of the pattern is next?

When the children return after the weekend or break, can they tell what symbols need to be placed on the calendar to represent the days not in school?

Making the Calendar

1. Create an "empty" calendar by drawing lines on a large piece of paper.

2. Cut squares the same size as the boxes drawn on the calendar.

3. Each month, have the children create the symbols you will use for the pattern for that month.

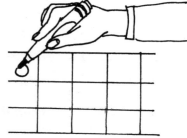

Ideas for Symbols:

Apple prints: red and green

(Pattern example: Red, red, green, green)

Sponge-print geometric shapes: triangle, rectangle, square, circle

(Pattern example: Triangle, triangle, rectangle)

Finger paint squares using whatever colors you want your students to identify

(Pattern example: Orange, green, yellow, orange, green, yellow)

Sometimes you can provide symbols that help develop vocabulary:

Tinfoil squares

(Pattern example: *Shiny, shiny, dull, shiny, shiny, dull* (or *not shiny*))

Paper shapes of different widths

(Pattern example: *Wide, narrow, wide, narrow,* or *fat, skinny, fat, skinny*)

Paper shapes in three sizes

(Pattern example: *large, medium, small, large, medium, small,* or *big, middle-sized, little.* Note: These could be paper cut-outs representing the bowls from *Goldilocks and the Three Bears.*)

How Many Days in School?

Materials:

- Circles to add together to make a caterpillar

  or

- Adding machine tape to unroll as the days go by

Lesson Focus:

- To observe a number line created by adding one more over and over again
- To see the pattern that emerges over time
- To predict what comes next

The Task:

Keep a cumulative record of the number of days of school by recording the days of school on paper circles representing the parts of a caterpillar that is getting longer and longer. Or write the numbers on an adding machine tape and put it up in the room to be unrolled as the days go by.

Interactions: Deepening the Understanding

Ask: ***"What do you think comes next?"***

Observations: Watching the Growth

Some children will be able to predict what comes next. Others will see that numbers go on and on, and there's always one more.

Rhythmic Patterns

Materials:

- None

Lesson Focus:

- To introduce the concept of pattern
- To provide ongoing practice with a variety of patterns of varying levels of difficulty

The Task:

Spending three or four minutes at a time, begin a rhythmic pattern and ask the children to join in as soon as they know what the pattern is.

Begin working with these patterns on the first day of school and continue throughout the year. Do not expect or wait for all the children to master a pattern before moving on to others. All the children will improve over time if given many, varied experiences. Provide for individual needs by interspersing simple patterns with more complicated patterns.

Act out a rhythmic pattern and ask the children to join you after a few motions. For example: *Clap, clap, slap, slap* (legs); *clap, clap, slap, slap* ...

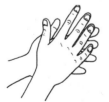 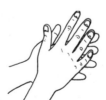  

Continue the pattern for at least 30 to 60 seconds. There is something very satisfying about the repetition of the rhythm. For children who are having difficulty, many repetitions of the pattern will help them begin to feel the pattern within themselves, even if they cannot yet act it out perfectly.

Encourage children to act out each pattern and say it along with you as soon it becomes obvious to them. Continue to direct the children so that no child feels put on the spot or doesn't know what to do.

Interactions: Deepening the Understanding

Do not single out individual children who need special help. Children who are having difficulty with a pattern may be less inclined to try it if their unsuccessful attempts are given attention.

When children have difficulty, teachers are often tempted to slow the pattern down. This actually makes the pattern more difficult to "feel." Keep the pattern moving at the rate of a child's heartbeat, about 90 times a minute.

Observations: Watching the Growth

Are the children able to join in with some measure of success?

What kinds of patterns and motions are easiest for them to do?

What happens if you keep the pattern going for a relatively long period of time? Are the children better able to feel the pattern after a few repetitions?

Over time, observe what happens when you vary the same pattern slightly in any of several ways. For example:

Verbalize one, but not all, of the motions as you act out the pattern:
Do: *Clap, clap, slap; clap, clap, slap …*
Say: **"* * slap; * * slap …"**

Verbalize all of the motions:
Do: *Clap, clap, slap; clap, clap, slap …*
Say: **"Clap, clap, slap; clap, clap, slap …"**

Add a variation to one part of the pattern:
Say: ***"This time let's nod our heads when we slap."***

Examples of other patterns you can introduce on succeeding days:

Clap, slap; clap, slap; …

Stamp, stamp (right foot), stamp (left), stamp (right); stamp, stamp (right foot), stamp (left), stamp (right); …

Nod, nod, clap; nod, nod, clap; …

(Touch the parts of the body as you say them.)

Clap, slap shoulders; clap, slap shoulders; …

Nose, shoulders, shoulders; nose, shoulders, shoulders; …

Nose, nose, nose, jump; nose, nose, nose, jump; …

Slap (legs), slap, clap, clap; slap, slap, clap, clap; …

Chin, shoulders, chin, knees; chin, shoulders, chin, knees; …

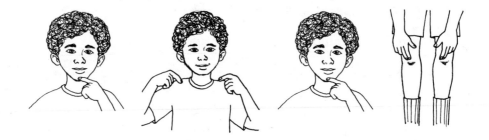

Developing Math Concepts in Pre-kindergarten

Weather Graph

Materials:

- Poster board or butcher paper
- Markers

Lesson Focus:

- To keep track of the weather on a daily basis
- To notice what kinds of weather are most common

The Task:

Have the children decide what kind of weather needs to be added to the weather graph. Then, you or a student draws a symbol or colors in a square on the graph. Over time, by observing the graph, the children will find out which kind of weather is most common.

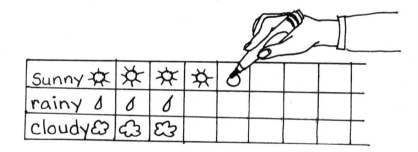

Interactions: Deepening the Understanding

Ask:

"What kind of weather is it right now? Is it raining, cloudy, or sunny? What weather happens most?"

Observations: Watching the Growth

Some children will recognize that the chart represents the weather they experienced during the preceding days and will be able to describe what kind of weather is common in the area for that time of year. Others will focus on the daily weather event itself.

Counting Poems and Finger Plays

Materials:

- A collection of counting books and poems

Lesson Focus:

- To reinforce the counting sequence through songs and poems
- To practice counting forward
- To experience the concepts of "one more" and "one less"

The Task:

Present a counting song, poem, or counting book, and have the children join in to practice the counting sequence. You may want to print these songs and poems on a poster board as part of a shared reading experience. For example:

Pit a Pat

Pit a pat
What was that?
Two small raindrops on my hat.
One, two.

(Children thump, thump their fingers on their heads to imitate the rain.)

Pit a pat
What was that?
Three small raindrops on my hat.
One, two, three.

(Children thump, thump, thump their fingers on their heads to imitate the rain.)

(And so on.)

—Author unknown

Observations: Watching the Growth

Make note of who is able to join in and who is not.

Developing Math Concepts in Pre-kindergarten

Examples of Counting Songs and Poems:

I Caught a Hare Alive

1, 2, 3, 4, 5,
I caught a hare alive;
6, 7, 8, 9, 10,
I let him go again.

—Author unknown

One, Two, Buckle My Shoe

One, two, buckle my shoe.
Three, four, shut the door.
Five, six, pick up sticks.
Seven, eight, lay them straight.
Nine, ten, a big fat hen.

—Author unknown

Five Little Snowmen

(Gives practice counting backward)

Five little snowmen fat,
(c   c   c   e   e   g)
Each with a funny hat (point to head)
(a   a   a c a   g)
Out came the sun and melted one.
(c   c   c c   b   a g g   e)
What a sad thing was that!
(f   f   f   e   d   c)
DOWN, DOWN, DOWN!
(c   a   c)
(The fifth snowman, exaggeratedly slowly, melts to the floor)

Four little snowmen fat,
Each with a funny hat (point to head)
Out came the sun and melted one.
What a sad thing was that!
DOWN, DOWN, DOWN!
(The fourth snowman, exaggeratedly slowly, melts to the floor.)

(Continue with three and two snowmen, ending with:)

One little snowman fat,
(S)he with a funny hat (point to head)
Out came the sun and melted one.
What a sad thing was that!
DOWN, DOWN, DOWN!

—Author unknown

Feel the Beat

Materials:

- None

Lesson Focus:

- To increase children's ability to count by rote beyond what they already know while doing a variety of rhythmic motions as they count

The Task:

Demonstrate the motions you want the children to make as they count, and invite them to join in. Spend just 2–3 minutes at a time 3–5 times a week. You do not need to spend a lot of time focusing on these short sequences.

Be sure you are in tune with those children who need the most practice and are not misled by loud voices from children who are already confident in doing the sequence.

For example, say, "We are going to practice counting to 4, and this is how we are going to do it." (Demonstrate the motions, emphasizing the final number, and have the children join in, repeating several times. For example, act out: March, march, march, arms up, as you say, "One, two, three, FOUR!")

Interactions: Deepening the Understanding

Your main focus in this activity will be on helping those children who are not yet able to count to 10. The activity is designed, however, to be of value also to those children who already know the counting sequence, since it develops rhythmic and motor skills in children.

Decide on how long to practice according to the needs of the children who need the most help with the sequence. For example, if you have just two or three children who need to count to three, take this time to respond to their needs without singling them out. All the children will have fun and improve their physical coordination and rhythm by doing these rhythm sequences.

Observations: Watching the Growth

Watch those children who need the most counting practice. Adapt the motions and sequences to meet their needs.

Examples of Actions:

Do: *Box (punch the air), box, CLAP …*
Say: ***"One, two, THREE!"***

Do: *Right leg up, left leg up, right leg up, left leg up, SQUAT …*
Say: ***"One, two, three, four, FIVE!"***

Do: *Jump, jump, jump, jump, TURN (quarter turn) …*
Say: ***"One, two, three, four, FIVE!"***

Do: *Stamp, stamp, stamp, KICK; stamp, stamp, stamp, KICK …*
Say: ***"One, two, three, FOUR; five, six, seven, EIGHT!"***

Examples of Rhythms:

It will be interesting for the children if you vary the cadence and rhythm of these motions. Some possibilities are:

Rhythm #1:
"ONE, TWO, three, four, five …"
Example: *STAMP, STAMP, twist, twist, twist …*

Rhythm #1:
One, two, three, four, FIVE …
Example: *Twist, twist, twist, twist, STAMP…*

Rhythm #2:
"One, two, THREE; four, five, SIX …"
Example: *Twist, twist, TWIST; twist, twist, ARMS OUT …*

Counting Jars

Materials:

- Various jars and other clear containers
- Sets of objects used to fill the jars (should be in various sizes: peanuts, pecans, pom-poms, blocks, ping-pong balls, and so forth)

Lesson Focus:

- To count along with the teacher using a rote sequence greater than the child is able to count alone
- To see modeling of one-to-one counting
- To use counting to find out how many
- To develop a sense of quantity and relationships between numbers

The Task:

Have the children count along with you to see how many objects (walnuts or cubes or peanuts or Ping-Pong balls) fit in an empty jar. For example, present an empty clear container. Say:

"I wonder how many walnuts we can put in this jar. How many do you think will fit? Let's count and see."

To focus their attention, you may ask them to make a guess first. You will see young children guessing any number they happen to know. The numbers are just words to them at this point, so don't refer to their guesses after the counting is completed.

Variations:

Jar Full of Objects

Present a container already filled with objects and have the children guess how many are in the jar. Count as you empty the jar.

Same Objects, Different Container

After determining the number of objects that fill a particular container, present a larger or smaller container and see whether the number is more than, less than, or the same as the first container. After some experiences, ask the children to tell what they think will happen before actually determining the amount.

Same Container, Different Objects

After determining the number of objects that fill a particular container, present different objects and see if the container will hold more than, less than, or the same as when filled with the objects used previously.

Interactions: Deepening the Understanding

Vary the pace of the counting so that the children have to pay attention. If they begin saying more than one number per object, stop and ask:

"Are we counting carefully? Did I put two walnuts in the jar or one walnut? Make sure we are counting what I put in the jar. One ... two ... three ... four ... five ..."

When you finish counting, ask: *"How many did we count?"*

Observations: Watching the Growth

Notice which children count with you as you count objects and which children are just saying the numbers or rushing ahead.

Notice who remembers the number counted.

Alike and Different

Materials:

- Various items available in the classroom: 6–7 should have attributes that are alike in some way and 3–4 should be different

Lesson Focus: To notice attributes that belong together

The Task:

Present the children with several objects and ask them to try to figure out what is alike about the objects and what is different.

Place a group of objects out of the children's view. Show them two items that are alike and ask them to tell what they think makes the two objects alike. Put the objects that go together in one pile and those that do not in another pile.

After several items have been sorted, discuss what makes the objects in the group alike. To help the children understand the activity, the attributes for the first experience should be fairly obvious. Later, when presenting additional experiences, the attributes can pose more of a challenge.

For example, bring out a red triangle and a red circle, and ask: **"Who would like to say how they think these shapes are alike?"**

Say: **"I am going to put the ones that are alike in this pile."**

Next, bring out a yellow circle and say: **"This one is different from the other ones."**

Say: **"I am going to put all the ones that are different over here."**

Bring out another item and say: **"Now I am going to show you something else."**

Developing Math Concepts in Pre-kindergarten

Say: ***"Thumbs down if you think it goes in this pile. If you are not sure, point your thumb sideways."*** After the children have had a chance to decide, put the item in the correct pile.

Show the children a red square.

After they have had a chance to decide, put the item in the correct pile.

Continue with several other items. When all the items have been sorted, discuss what makes the objects in each group alike.

Interactions: Deepening the Understanding

If the children are having difficulty focusing on what makes one item like another, ask: ***"What do you notice about the color?"*** or ***"Are these the same shape? Are they the same size? Are they the same color?"***

Over time, present less obvious attributes—for example:

- Pictures of farm animals and animals that are not found on a farm

- Objects with wheels and without wheels

- Items that are all the same color but different shapes

Observations: Watching the Growth

When comparing two objects, are the children able to attend to one attribute at a time, such as color or shape, or do they identify several attributes?

Are they able to tell which one thing is the same about two objects?

Are they able to label a group of objects with the same attribute? ("These are all the red ones.")

Descriptions

Materials:

- Classroom or household items that have similar uses but look different (for example, a saucepan and a frying pan, two types of balls, a fork and a spoon, a shoe and a sandal)

Lesson Focus:

- To focus on attributes and describe them
- To identify attributes that are the same

The Task:

Have the children get ready to sort by describing the attributes of various objects that are similar in use but look different. Then, have them determine what is the same and what is different about the objects.

Ask the children to tell you something about a particular object. As the children share attributes, you may want to write their descriptive words on the board. For example, ask: **"Who can tell me something about this mug?"**

"It's white."

"It's white, but it has some blue and red on it, too."

"It has a handle."

"You can put stuff in it."

Ask: **"Who can describe something about this glass?"**

"It's white."

"No, it isn't. It's called clear."

"It's glass."

"The sides are straight."

Ask: **"Who can tell us something that is the same about the glass and the mug?"**

"They both hold stuff."

"You can drink out of them."

Ask: **"What's different?"**

"They're not the same color."

"One has a handle; the other one doesn't."

Interactions: Deepening the Understanding

Ask questions that focus the children's attention on the many different attributes of the objects being used. Also, help them identify the attributes that are the same and those that are different. For example, say:

"You said this glass was something to drink out of. Is the mug something to drink out of?"

"Yes, the glass and the mug are alike because you can drink out of them."

Say: **"You said the glass is clear. Is the mug clear?"**

"No, the mug is not clear. The mug and the glass are different because one is clear and the other is not clear."

Observations: Watching the Growth

Do the children have words to describe the objects?

Are they able to identify attributes that are the same and attributes that are different?

People Patterns

Materials:

- None

Lesson Focus:

- To notice and describe a pattern set up by the teacher
- To extend the pattern and predict what will be next

The Task:

Have the children sit in a circle on the floor. Direct the first few children to pose in different positions to create a pattern. Point to each child in order. Say, for example:

"Hands up, hands up, hands down, hands down, hands up, hands up, hands down, hands down ..."

As soon as the pattern becomes obvious, have the class say the pattern with you. Continue the pattern, including more children as you go around the circle. Continue to direct the children without putting anyone on the spot. Keep this a group experience where everyone knows what to do and is comfortable when it is time for their turn.

Like the Rhythmic Patterns activity, People Patterns can be an ongoing part of the year, since the level of difficulty can be raised as the children's ability to work with patterns grows.

The following are examples of other patterns you can do on succeeding days. You can also have the children participate in making suggestions.

Turn, turn, straight; turn, turn, straight …

Sit, sit, stand; sit, sit, stand …

Bend, straight, bend, turn; bend, straight, bend, turn …

Interactions: Deepening the Understanding:

Ask the group as a whole: ***"What do you think comes next?"***

Observations: Watching the Growth:

Are the children able to demonstrate which step of the pattern they are to do when it's their turn? Do they do this with ease or difficulty?

Do they anticipate the motion before being directed to do so, or do they need to be told what motion they are to do?

Patterns in the Environment

Materials:

- Various, according to the patterns being explored

Lesson Focus:

- To notice patterns that appear in the environment

The Task:

Help the children see that patterns are everywhere by pointing out patterns in the classroom. "I see a pattern" can become an invitation from you (or from one of the children) to look for a pattern. For example, you could say:

"I see a pattern when I look at John's shirt. Does anyone else see the pattern?"

Once children have had an opportunity to look for the pattern, you or a child can describe what you see:

> "Red stripe, white stripe, black stripe, red stripe, white stripe, black stripe …"

Patterns commonly can be found on clothing. You and your students may also discover patterns in the design of food packaging, wall coverings, and book covers.

You may also find some perhaps less obvious patterns in:

Fences (tall, short; tall, short, …)

Gardens or hedges (bush, flowers; bush, flowers; …)

Buildings (wall, window; wall, window; …)

Dish designs (flower, diamond; flower, diamond; …)

Once children discover other real-world patterns, they will be eager to share them with the class.

Interactions: Deepening the Understanding

Notice patterns as they arise throughout the school day. Say: ***"I see a pattern. Does anyone else see the pattern? Let's say the pattern together."***

Observations: Watching the Growth

Do the children find patterns on their own and report them to the class?

Do they recognize the patterns shared by others?

Real Graphs/Name Graphs

Materials:

- Various classroom objects: Books, blocks, snacks, shoes, name or picture cards, paper plate with the child's name on it, and so on

Lesson Focus:

- To organize information on a graph
- To make comparisons

The Task:

Children should have a variety of data collection experiences throughout the year. Watch for opportunities that arise spontaneously from actual classroom events that lend themselves to organizing information in a graph. The children should experience a variety of types of graphs: real graphs, name and picture graphs, and weather graphs. Pose a question and have them answer by making a graph.

For example, in creating a real graph, ask:

"What kind of fruit did you choose for snack today—a banana or an orange?"

Indicate where on the graph the children are to place their fruit and then have them look for the relationships revealed by the graph.

Noticing and describing relationships is the purpose of graphs and an opportunity for developing ideas of more and less. After the children have placed their information on the graph, point out the columns and say: **"Look at our graph. Can we tell without counting which column has more—the oranges or the bananas? Do we know which has less?"**

Interactions: Deepening the Understanding

Say: **"Let's count together and see how many."**

"What else do you notice?"

Observations: Watching the Growth

In observing the information displayed on the graphs:

Do the children match one to one?

Can they tell which category has more and which has less or fewer?

Real Graphs

The following are examples of the kinds of questions that come up on a daily basis and can be answered by organizing the objects themselves into real graphs to make comparisons.

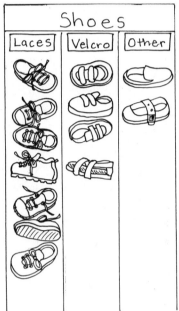

"Do your shoes have laces or Velcro or something else?"

"Did you bring a red apple, a green apple, or a red and green apple to make applesauce?"

"What kinds of vegetables did we bring for our Stone Soup?"

"Did we collect more yellow leaves or brown leaves today on our walk?"

"Are there more of this kind of shells on the science table or this kind?"

"Is your macaroni necklace red and blue or orange and green?"

"Are there more children who picked oranges or more children who picked apples for a snack?"

Name Graphs

The children answer a question by placing their name card or picture of themselves on a graph.

"Did you walk or ride to school today?"

The following are examples of questions that can be answered with a "yes" or "no" and organized on a name or picture graph:

"Do you have a sister?" "Do you have a brother?"

"Are you wearing green today?"

"Do you have a J in your name?"

"Do you have an A in your name?"

"Is your foot longer than, shorter than, or the same as this line?"

The following are examples of questions that ask children to make decisions or choices:

"Should we read this book or this book?"

"What game do you want to play when we go outside—this one or this one?"

"Do you want to paint your box building red or brown?"

Geometry Walks: Looking for Shapes

Materials:

• Objects found in the classroom or schoolyard

Lesson Focus:

• To observe shapes in the natural environment

The Task:

Take the children on walks outdoors or around the classroom in order to discover and describe shapes in the world. Ask:

"What circles do you see? What rectangles? What triangles?"

"What is the shape of the table? What else can you find that is that shape?"

"What shapes do you see when you look closely at the door? The windows? Our dishes?"

The geometry walk is intended to help children become more observant about the shapes and patterns that exist in the world around them. The children will notice:

"The clock is a circle."
"The table is a square."

We can also help children see the shapes within larger objects. For example, say: ***"If we look closely at a car, what shapes can we see? Do you see any circles?"***

Interactions: Deepening the Understanding

Help the children tune in to the repetitive forms in nature. (It's all right to look in books as well as in the world around us.)

"Where can we find circles inside of circles?"

"What can you find that has a circle in the middle and rays or spokes coming out from that circle?"

"What does a spiral look like? How many different spirals can we find?"

"What do we notice when we look closely at the veins in leaves? How are they alike? How are they different?"

Observations: Watching the Growth

What do the children pay attention to?

Are they enthusiastic? Is there a sense of discovery?

Can they see the smaller shapes that make up a particular structure or form, or do they focus more on the basic shapes and forms?

Do they notice forms in nature that have similarities? What attributes do they notice?

Are any interested in representing their discoveries on paper?

What's the Same as Me?

Materials:

- Objects found in the classroom
- Paper and markers (optional)

Lesson Focus:

- To begin to notice size and length

The Task:

Have the children make direct comparisons by searching for objects in the room that are about the same length as one of their body measurements. After exploring, have the children share their discoveries with the group.

For example, say:

"We are going to look around and see if we can find things in the room that are as long as our arm."

Choose a volunteer to model what the children are going to do. Say:

"Let's look at Johnny's arm first. Do you see anything in the room that might be the same length as Johnny's arm?"

Try out a couple of suggestions that the children make and determine together whether the object is the same length or not. ("This book is the same length as my arm.")

Next, give all the children a few minutes to find and measure objects with their arms to see if they can find any they believe are the same. Then, have those who wish to share tell what they found.

Over time, provide additional opportunities for the children to work with other measurements by finding things about the same length as their foot, hand, hand span, leg, and so forth.

Interactions: Deepening the Understanding

Looking for things that are the same length as a part of their body helps children focus on the property of length. As they look for objects that match, they also work with the idea of precision, since they have to decide how close is "close enough" to be considered the same length.

Ask:

"How did you decide if it was close enough to call it 'the same'?"

"Did everybody else find the same things as you did when they measured with their arm? Why or why not?"

Observations: Watching the Growth

Do the children try measuring things that seem unlikely to be the length they are looking for, or are they discriminating, measuring only those things that are fairly close?

Are they misled by a lack of perspective; that is, do they need to get really close to an object before they can tell if they have found something that might be the same?

Do they become more discriminating with experience?

How exact do they feel they need to be? Does their idea of precision change over time?

Is It Longer or Shorter?

Materials:

- Objects found in the classroom

Lesson Focus:

- To make direct comparisons by finding objects in the classroom that are either a little longer or a little shorter than a particular body measurement

The Task:

Ask the children to look around the room for objects that might be a little longer or a little shorter than the part of their body they are comparing. Let one or two children check out their ideas while the rest of the class observes. Or let all the children search for objects for a few minutes. Have the children share what they found.

Provide additional opportunities, over time, for the children to work with measurement by finding out if objects in the classroom are a little longer or shorter than other parts of their body (foot, hand, hand span, leg, and so on).

Interactions: Deepening the Understanding

While the children work, give them opportunities to use comparative language by asking: ***"What did you find out?"***

You will notice that expressing comparisons can be difficult for children. If a child is looking for things that are longer than their hand, for example, they will be able to say,

"The book is longer than my hand."

However, it will not be as easy for them to make the reverse comparison and say,

"My hand is shorter than the book."

You may need to model the language for them. For example, you might say:

"You told me that the book is longer than your hand. That means your hand is shorter than the book, too."

Observations: Watching the Growth

Are the children able to find objects that are both longer and shorter, or do they focus only on one relationship at a time?

Can they predict ahead of time what object they think it will be?

Are they able to describe what they found out?

Measuring with Strings

Materials:

- String cut to match the lengths of several objects in the room

Lesson Focus:

- To experience measuring by searching for objects that are the same length as various pieces of string
- To introduce the idea of a measuring tool

The Task:

Have the children use a length of string to find objects that are the same length as the string.

Variation:

Have the children measure objects and report whether the object is longer than, shorter than, or the same as the one being compared.

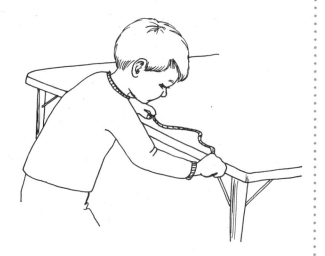

Interactions: Deepening the Understanding

Ask: ***"Is it easier to measure with your arm or a string?"***

Observations: Watching the Growth

Do the children try measuring things that seem unlikely to be the length they are looking for, or are they discriminating, measuring only those things that are fairly close to the length of their string?

Are they misled by a lack of perspective; that is, do they need to get really close to an object before they can tell if they have found something that might be the same length?

How close do they need to be before they are sure they have found the right object?

Are they able to comment on what they have learned?

Setting 3: Teacher-Directed Small-Group Time

Keep the group lessons short so that you will still have time after the lesson of the day to be involved with your students and observe them as they work with the materials and activities independently.

It is important to observe your students very carefully during small-group work. This is your chance to see what the children are thinking. They will respond in many different ways. Watch for that and delight in it. As you watch the children, you will be able to adapt the level of difficulty (the size of the numbers, the complexity of the patterns, the nature of the attributes used for sorting) according to their responses.

It is very important to recognize that the small-group experiences need to be repeated numerous times. Children need many experiences with the same idea to develop the understanding and facility we are aiming for.

Teacher-directed Small-group Activities:

Number:

Ding!, p. 100

Dump It Out, p. 101

Shake Them Up, p. 102

Making Buildings, p. 103

"Let's Pretend" Stories, p. 105

Acting Out Songs and Poems, p. 106

I Changed My Mind, p. 110

Roll Again!, p. 112

One More, p. 114

Fingers and Cubes, p. 116

Taking Turns, p. 117

What's Hiding?, p. 118

How Many Do You See? Using Dot Cards, p. 120

How Many Do You See? Using Number Sets, p. 121

Can You Find It?, p. 122

Concentration, p. 123

Sorting and Patterns:

Sorting Attribute Blocks, p. 124

Sorting Collections with the Teacher, p. 125

Sorting Shapes, p. 126

Sorting Blocks, p. 127

Sorting Boxes, p. 128

Geometry and Measurement:

Copying Designs on a Geoboard, p. 130

Grab Bag Geometry, p. 131

Measuring Containers, p. 133

Comparing Objects Using Scales, p. 134

Is It Heavier or Lighter?, p. 135

Ding!

Materials:

- Xylophone
- Counters: Unifix cubes, wooden cubes, tiles, or collections (Have all the children use the same type of counters.)

Lesson Focus:

- To focus the children's attention on one-to-one correspondence as you count with them
- To emphasize the counting motion by using a xylophone

The Task:

Have the children slide counters one at a time toward themselves as they count together with you. Have them repeat counting to a particular number several times. One group of children may need practice counting to four, while another group of children may need practice counting to six.

For example, say: ***"We are going to practice counting to four."***

Play a note on the xylophone on each count. Together, you and the group count, "One, two, three, four."

After the number has been counted, say ***"Check,"*** and have the children recount their counters.

Run the stick across the xylophone as a signal to push the counters back into the pile. Repeat several times.

Interactions: Deepening the Understanding

Say: ***"Let's practice counting. I will play the xylophone and we will count together."***

Observations: Watching the Growth

Are the children able to tell the number of counters they have when asked to check?

Dump It Out

Materials:

- Margarine tub for each child
- Counters: Unifix cubes, wooden cubes, tiles, or collections (Have all the children use the same type of counters.)

Lesson Focus:

- To help the children focus on one-to-one correspondence as they count with you

The Task:

Have the children each drop counters into their margarine tub as they practice counting to a designated number. You or a student can say, "Dump it out," and the counters are dumped and recounted. For example, say:

"Let's count together."

"One, two, three, four."

"Dump it out! Let's count again."

"One, two, three, four."

Repeat several times, choosing different numbers.

Interactions: Deepening the Understanding

Vary the pace of counting to help the children focus on saying one number for each counter.

Observations: Watching the Growth

Are the children able to keep track of the objects when asked to count again?

Are they able to switch from counting one group to counting another?

Shake Them Up

Materials:

*   Counters: Unifix cubes, wooden cubes, tiles, or collections (Have all the children use the same type of counters.)
*   Paper bag

Lesson Focus:

*   To practice counting objects
*   To experience constancy or conservation of number

The Task:

Have the children count with you (or ask one child to count) as you put the counters into the bag. Have one child shake the bag, and then have each child in the group whisper in your ear how many counters they think are in the bag. Dump the counters out and check the number. Put a different number of counters in the bag and repeat the activity. Do this several times.

Interactions: Deepening the Understanding

This activity gives children practice in counting, but it also gives them something else to find out. Ask: ***"Does the number change when it is inside a bag? What happens if you shake up the bag?"***

By asking the children to whisper their ideas, you allow them to share their thinking with you without being influenced by the other children's thinking. Notice which children are not sure what number is in the bag even after many experiences. Those who aren't sure how many counters there are when the counters are hidden have not yet learned conservation of number.

Observations: Watching the Growth

Notice which children are convinced from the very beginning that the number does not change and which children's ideas change after several experiences.

Remember, children must come to this understanding on their own. Do not be disheartened if some still do not know that the number will be the same even after several countings. The real focus of this activity is counting practice, and they are getting plenty of that! The reason for hiding the counters in the bag is to provide the opportunity to ponder what is happening if they are ready to think about it. Be very sure not to teach them "the right answer." We do not want children to try to please the teacher by saying words they don't yet understand.

Making Buildings

Materials:

- Unifix cubes

Lesson Focus:

- To focus on the counting sequence by adding one cube at a time

The Task:

Direct the children in making Unifix buildings by adding one cube at a time, having them count as high as appropriate to their needs.

Choose one cube. Ask: ***"How many do we have?"***

"One."

Pick up another cube. Ask: ***"How many now? Let's count."*** Emphasize the word *two* to focus on the total. Continue emphasizing the last number as you go.

"One, *two.*"

Take one more. Ask: ***"How many now? Let's count?"***

"One, two, *three.*"

Take one more. Ask: **"How many now?"**

"One, two, three, *four*."

When the children have counted as high as you wanted them to, have them make a Unifix building by stacking the cubes. Build several stacks with them to give them a lot of practice with the sequence they are working on.

Interactions: Deepening the Understanding

At some later date, when the children are able to count easily, ask them to tell how many without counting, if they can.

Observations: Watching the Growth

Are the children able to keep track as they count, or do they lose count?

Are they able to predict how many there will be when one more is added? If so, to what size number?

"Let's Pretend" Stories

Materials:

- Colored construction paper to represent different environments: ocean (blue), grass (green), cave (black), and so on (see page 28)
- Unifix cubes sorted by color or tiles or plain wooden cubes (Have all the children use the same type of counters.)

Lesson Focus:

- To practice counting by acting out counting stories
- To practice counting by adding one more

The Task:

Tell stories for the children to act out. For example, say:

"Today we are going to work with the 'let's pretend' ocean. It is a stormy day on the ocean. The waves are high and it is starting to rain. There are four ships on the ocean. Show me the ships. Three of the ships decided to go back to the harbor ..." And so on.

Or: *"There are five big whales in the ocean today, too. Show me the whales."*

Interactions: Deepening the Understanding

You can involve the children in what is just more counting practice simply by engaging their imaginations. Let the children be involved in deciding what might be in the ocean (or cave or playground) and what might be happening there. Then you can use their ideas to tell the story while making sure the numbers given are appropriate ones for the group to be counting.

Vary the level of difficulty by changing the size of the numbers. Your purpose here is not finding out how many are left or how many there are altogether, but some children will be figuring that out, too.

Observations: Watching the Growth

Are the children engaged?

Are they accurate when they count?

Acting Out Songs and Poems

Materials:

- Colored construction paper to represent different environments (see list of materials under each song or poem)
- Counters: Unifix cubes sorted by color or tiles or plain wooden cubes (Have all the children use the same type of counters.)

Lesson Focus:

- To practice counting by acting out counting songs and poems

The Task:

Have the children act out various songs and poems using colored construction paper and counters. Although children love to be the birds or the kittens and act out the poems in this activity, there are advantages to using the "Let's Pretend" materials listed. Each child is involved and must actually do some counting. Choose from the following poems or others that you know. It is not necessary to do all of them with every student.

BEEHIVE

One little bee flies round and round;
One little bee can make this sound
Zzzzzzzzzzz

Two little bees fly round and round;
Two little bees can make this sound
Zzzzzzzzzzz

Three little bees fly round and round;
Three little bees can make this sound
Zzzzzzzzzzzzz

Four little bees fly round and round;
Four little bees can make this sound
Zzzzzzzzzzzzz

Five little bees fly far away
But they will come again someday
Zzzzzzzzzzz (softer and softer until it fades away)

—Author Unknown

Materials:

- Blue construction paper (sky)
- Counters to represent bees (Yellow Unifix cubes work well, as they can be placed on the children's fingers and the children can make them fly.)

Variations:

Ask the children, "What else could fly around? What sound could it make?" Use their ideas to change the poem. Some examples might be:

One little airplane flies round and round;
One little airplane can make this sound.
Rrrrrrrrrrr

One little crow flies round and round;
One little crow can make this sound.
Ca-a-a-w

THE BEE

(Make sure the children touch each bee as they make the buzzing sound. This will help reinforce one-to-one counting.)

There was a bee
Upon a wall.
"Buzz," said he,
And that was all.

There were two bees
Upon a wall.
"Buzz, buzz," said they,
And that was all.

There were three bees
Upon a wall.
"Buzz, buzz, buzz," said they,
And that was all.

There were four bees
Upon a wall.
"Buzz, buzz, buzz, buzz," said they,
And that was all.

There were five bees
Upon a wall.
"Buzz, buzz, buzz, buzz, buzz," said they,
And that was all.

—Author unknown

Materials:

- Brown or tan construction paper (wall)

- Cubes or counters to represent bees

Variations:

Ask the children, "What else could be on the wall? What sounds could it make?" For example,

There was a cat
Upon a wall.
"Meow," said he,
And that was all.
(And so on…)

There was a mouse
Upon a wall.
"Squeak," said he,
And that was all.
(And so on…)

COUNTING KITTENS

One kitten on the floor,
Two kittens by the door.

Three kittens roll a ball,
Four kittens in the hall.

Five kittens gently purr,
Six kittens wash their fur.

Seven kittens in the house,
Eight kittens smell a mouse.

Nine kittens softly creep,
Ten kittens sound asleep.

   —Author unknown

Materials:
- Green construction paper (back-yard)
- Cubes or pompoms to represent kittens

LITTLE BIRDS

One little bird with lovely feathers blue
Sat beside another one. Then there were two.

Two little birds singing in the tree;
Another came to join them. Then there were three.

Three little birds wishing there were more;
Along came another bird. Then there were four.

Four little birds, glad to be alive,
Found a lonely friend. Then there were five.

Five little birds just as happy as can be,
Five little birds singing songs for you and me.

—Author unknown

Materials:
- Blue construction paper (sky) or green paper (grass, tree, or nest)
- Cubes to represent birds

FIVE LITTLE SEASHELLS

(The children will move the shells from the shore or beach to the ocean according to the poem. Count along together with the children as long as needed.)

Five little seashells, washed up on the shore,
Swish! went the waves, and then there were four.

Four little seashells, as pretty as can be,
Swish! went the waves, and then there were three

Three little seashells, all shiny and new,
Swish! went the waves, and then there were two.

Two little seashells, lying in the sun,
Swish! went the waves, and then there was one.

One little seashell, having lots of fun,
Swish! went the waves. Now there are none!

—Author unknown

Materials:
- Blue construction paper (ocean)
- Sand paper (beach)
- Cubes to represent shells or real shells

I Changed My Mind

Materials:

- Counters: Unifix cubes, wooden cubes, tiles, or collections
- Organizers: Egg cartons, six-pack plastic rings, Working-Space Paper (BLM #1), Ten Frames (BLM #2), Ten Strips (BLM #3)

(Have all the children use the same type of counters and organizers.)

Lesson Focus:

- To practice counting objects
- To provide opportunities to see relationships between numbers

The Task: Level 1: Counting

Call out numbers that you want the children to build on their organizers. For example, say: ***"Show me four."***

(The children put out four counters.)

Say: ***"Oh! I changed my mind. I wanted seven."***

(The children put out seven counters.)

Say: ***"I changed my mind again! Now I want six."*** And so on.

Do not have the children arrange their counters in any particular way on the organizer. You want them to see the same number arranged in a variety of ways. On different days, change the type of organizer you use with the children so they can see numbers organized in many different ways. For example:

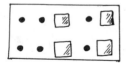 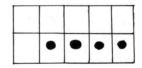 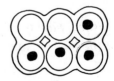

Developing Math Concepts in Pre-kindergarten

Interactions: Deepening the Understanding

Say: *"Let's look at Linda's organizer. Did she put her counters in the same way as you or a different way? What about Taylor?"*

Observations: Watching the Growth

This very simple yet powerful activity can be done over and over again with your students. It can serve as a way for children to practice counting, but it also has the potential for helping them see the relationships between numbers.

You will see the many different ways that children handle this activity. Most will take all the counters off their organizer each time before they build a new number. They see each number as something new and different and unrelated to the previous numbers they built. A few may count all the counters on their organizer and then add some on or take some off. On occasion, children may see that when they have four and the teacher says six, they need to get two more.

You can make this appropriate for different groups of children by changing the size of the numbers. You can also make it more difficult if you ask the children to build numbers with larger differences between them, like two and five, rather than numbers that are closer together, like six and seven.

The Task: Level 2: Counting and Numeral Recognition

Help children learn numerals and associate them with the amounts they represent. Instead of calling out the numbers that you want the children to build on their organizers, write the numerals as you say them. Sometimes, write the numeral without saying it and ask the children to read the numeral.

Interactions: Deepening the Understanding

Say: *"What do you think this numeral is? Can you show me?"*

Observations: Watching the Growth

Are the children associating the numerals with the amounts they represent?

Are they beginning to read the numerals as you write them instead of waiting for you to say the numeral?

Roll Again!

Materials:

- Counters: Unifix cubes, wooden cubes, tiles, or collections
- Organizer for each child: Egg cartons, six-pack plastic rings, Working-Space Paper (BLM #1), Ten Frames (BLM #2), or Ten Strips (BLM #3)
- Large dotted dice
- Large numbered dice

(Have all the children use the same type of counters and organizers.)

Lesson Focus:

Levels 1 and 2:

- To practice counting objects
- To see relationships between numbers
- To match the number of dots on the dice and the number of counters

Level 2:

- To recognize numerals
- To match the quantity with the corresponding numeral

The Task:

Have the children roll a die to determine the number they are to build on their organizer. If you want the children to work with small numbers, use a 0–4, 0–5, or 1–6 die. If you want them to work with larger numbers, use a 4–9 die.

Level 1:

You and/or the children roll a dotted die and tell the number that comes up. For example: The die shows 4 dots. Say: ***"How many dots? Show me with the counters."*** The children put out four counters.

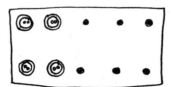 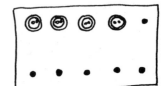

Developing Math Concepts in Pre-kindergarten

Say: **"Roll Again!"** The die shows 6 dots. The children put out six counters.

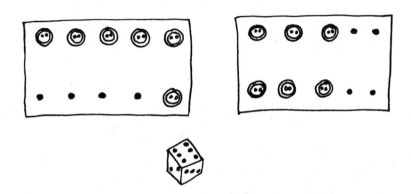

Say: **"I changed my mind again! Now I want eight."** And so forth.

Level 2:

Roll the numbered die to determine the number.

Interactions: Deepening the Understanding

Just as in the activity "I Changed My Mind," the children can look at other children's arrangements and decide if they look the same or different than their own.

For example, say:

"Let's look at what Joey did. Does his organizer look the same as yours? Let's count and see how many counters are on all our organizers."

Observations: Watching the Growth

Are the children able to determine the number by reading the dice? Are they able to keep track of the amount?

Level 1: Do any children recognize the dice arrangements without counting?

Level 2: Do they recognize the numerals before you read them aloud?

One More

Materials:

- Counters: Unifix cubes, wooden cubes, tiles, or collections
- Organizer for each child:
    - Egg cartons
    - Six-pack plastic rings
    - Working-Space Paper (BLM #1)
    - Ten Frames (BLM #2)
    - Ten Strips (BLM #3)

Lesson Focus:

- To practice counting objects
- To practice determining the quantity without counting after one more object is added

Variation:

- To practice determining the quantity without counting after one object is taken away

The Task:

Call out a number for the children to build in their organizer. Then say: "Put one more. How many now?"

For example, say: ***"Show me 4."***

Say: ***"Put one more. How many now?"***

"Let's start again. Show me 3."

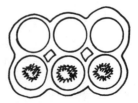

"Put one more. How many now?"

Repeat several times.

Variation:

If you have a group of children who know all the numbers to ten without counting, change the game by saying: *"Take one off. How many now?"*

Interactions: Deepening the Understanding

Some children will count even when they already know how many there will be when one is added. To encourage them to use what they know, say:

"Before we count, can you tell what number you think it will be?"

Ask the question as though you were just wondering, without expecting a correct answer. This way, if the children don't know the answer or are just guessing, they won't think they are supposed to try to figure it out.

Observations: Watching the Growth

Watch to see if the children need to recount to see how many objects they have after adding one to their organizer. Perhaps some of them will be able to recognize the small numbers without counting. Notice at what point in the sequence they start needing to count to tell how many.

Fingers and Cubes

Materials:

- Unifix cubes

Lesson Focus:

- To practice counting objects
- To develop flexibility and conservation of number by representing a number in a variety of ways

The Task:

Have the children work in pairs. When you call out a number, the first partner puts the appropriate number of cubes on the second partner's fingers on one or both hands.

Then, repeat the same number and have the first partner find another way to show that number on the second partner's fingers. This is repeated one more time before changing roles.

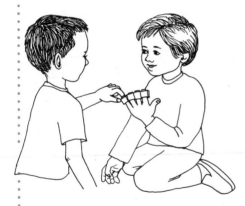

For example, say: ***"Show me four."*** The child puts four cubes on his partner's fingers.

Then say: ***"Show me another way to make four."***

The child changes the way the cubes are placed on the partner's fingers, putting two cubes on one hand and two cubes on the other hand.

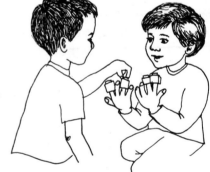

Repeat one one more time before the partners change roles.

Interactions: Deepening the Understanding

Say: **"You changed the cubes on your partner's fingers. How many do we have now? Are you sure?"**

Observations: Watching the Growth

Do the children know that the number doesn't change when they rearrange the cubes on their partner's fingers, or do they need to count to make sure?

Taking Turns

Materials:

- Counters
- Margarine tub or paper cups

Lesson Focus:

- To practice counting objects

The Task:

Have the children take turns putting counters into a container or taking them out. When each child in the group has had a turn, have them check to see how many ended up in the container. Then dump the counters out and start around the circle again.

For example, say: **"Paul, put in three."** All the children count with Paul as he puts in three. They continue to count with each classmate as they take their turn.

Other examples:
"Linda, put in two more."
"Peter, take out one."
"Xiong, put in four."
"Carlos, take out three."

Ask: **"How many do you think we have now?"** Have the children dump out the counters and count to see how many they have ended up with.

Interactions: Deepening the Understanding

This game allows you to make appropriate challenges for each child. If Peter has trouble with one-to-one counting, he is asked to put in one or take out one. Maybe Xiong can handle bigger numbers. But all are getting some practice as they count together to help their friends. The potential is there for some children to keep track as they go and to notice things like "We had three, we put in two, then we took out two, so we must still have three." You may encourage this by saying once in awhile, "How many do you think we have so far?"

Observations: Watching the Growth

Notice the level of confidence and the accuracy of the children's counting. Does this grow over time?

Are you able to ask them to add or take away increasingly larger numbers?

What's Hiding?

Materials:

- Counters
- 3 margarine tubs or paper cups

Lesson Focus:

- To practice counting objects
- To find out what happens to numbers when they are hidden from view and moved around

The Task:

Have the children count with you as you put groups of counters into three different margarine tubs. Tip the tubs over so that the counters are hiding beneath the tubs.

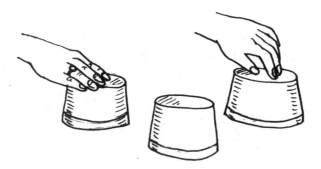

Move the tubs around and then have the children guess what number of counters is hiding under each tub. Lift and check after each guess.

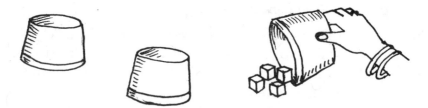

When all three tubs have been checked, rearrange them again. Repeat several times.

This is a chance to count (and count and count!) the same groups of counters over and over again, if your students need that practice. The children remain interested because there is an element of mystery, and they want to count again to see what has been hiding.

Interactions: Deepening the Understanding

When you lift the tubs to see the number of counters hiding, the children may recognize the number of objects in some groups without counting. Don't have them actually count the quantity unless some children are not sure and need to count. We want them to begin to trust what they know and not automatically count to check.

Observations: Watching the Growth

Hiding the counters also provides the opportunity to figure out something else. Some children will not realize that the number of objects stays the same as the number hidden originally. Other children might not be sure at first but will soon figure it out. There may be other children who are confident of the possibilities and are sure, by the end of the activity, of the number hidden because they have seen that it doesn't change.

How Many Do You See?
(Using Dot Cards)

Materials:

- Counters
- Dot Cards (BLMs #4–13)

Lesson Focus:

- To practice counting objects without touching them
- To match one to one
- To practice recognizing quantities of up to six without counting

The Task:

Display a dot card. Have the children take the number of counters they think they will need to cover the dots on the card. Next, put the card in front of one child and have him place his counters on the dots. Then, have the whole group count together to see how many. Repeat using another card.

After the children have become familiar with the cards, hold up one card and ask them to tell how many. Encourage the children to try to recognize the number of dots without counting when the cards show five or fewer dots. Some children will see a group of two or a group of three on a card but not be able to tell the total without counting.

Interactions: Deepening the Understanding

Ask: ***"How many dots do you think there might be on the card?"***

Observations: Watching the Growth

Are the children able to identify the number of dots without actually touching them to count?

Are they able to create groups of objects that match the dots?

What size numbers can they work with? Can they recognize groups of twos or threes without counting? Does the size of the numbers they can recognize increase over time?

For a related activity, see "Cover Them Up" (page 138).

Developing Math Concepts in Pre-kindergarten

How Many Do You See?
(Using Number Sets)

Materials:

- Number-set cards, portion cups, beads on a string, or other number-set materials (see pages 27–28)

Lesson Focus:

- To practice counting objects without touching each object

The Task:

Hold up the number-set cards one at a time and ask the children: ***"How many are on the card (in the portion cup, on this string, and so on)?"*** Then count together and see.

When working with number sets of five or less, let the children know that it is good when they can tell you how many without counting. Children sometimes think they are supposed to count every time, even when they can tell the quantity just by looking. You want to encourage the instant recognition of numbers to 4 or 5.

Interactions: Deepening the Understanding

Encourage the children to realize that they can tell how many are in a group without counting by asking: ***"How many objects (buttons, toothpicks, and so on) do you think there might be on the card?"***

Observations: Watching the Growth

Are the children able to identify the number of objects without actually touching them to count?

What size numbers can they work with? Can they recognize groups of twos or threes or fours without counting?

Does the size of the numbers they can recognize increase over time?

Can You Find It?

Materials:

- Number-set cards (see pages 27–28)

Lesson Focus:

- To find a matching quantity for a given number
- To practice counting objects that can't be moved
- To practice recognizing quantities up to six without counting
- To count quantities without touching each object

The Task:

Hand out 5–6 number-set cards to each child in the group. Hold up a number-set card and ask if anyone has a card with the same number as yours.

Interactions: Deepening the Understanding

Count with the children if they are having trouble determining the quantity without touching each object.

Observations: Watching the Growth

This activity gives the children a reason to count as well as some practice counting, even when they don't have a card with the same number as the teacher.

Notice the way they go about finding out how many are on your card. Can they tell right away? Do they appear to be counting just by looking, with heads bobbing, or do they need to actually touch the objects on the number-set cards?

Observe how they search to see if they have a number-set that is the same as yours. Do they count the objects on every card even though it couldn't possibly have the same number? Do any know instantly?

Can they eliminate some cards and just check the ones that look like they have close to the same number?

Developing Math Concepts in Pre-kindergarten

Concentration

Materials:

- Number-set materials previously sorted into sets of two for each number being worked with (for example, two cards with 3, two cards with 4, and so forth) (see pages 27–28)

- Margarine tubs, sheets of construction paper, or other materials to cover the number-set materials (If you are using portion cups with beans or pompoms glued inside them, they can just be turned over.)

Lesson Focus:

- To practice recognizing quantities up to six without counting
- To find a matching pairs of a given number
- To practice counting objects that can't be moved to keep track

The Task:

Set up the number-set cards or cups in three rows of four and cover them or turn them over. Have the children take turns uncovering two items to see what number is hiding. If they uncover a matching pair, they remove them. If the pairs don't match, they cover them again and the next child gets a turn.

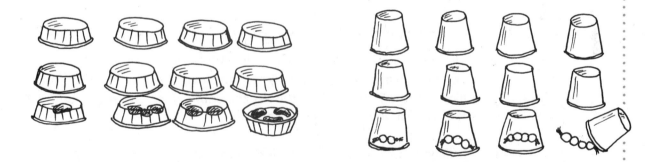

Interactions: Deepening the Understanding

Ask:

"Before you lift this, do you think you know what number is going to be underneath?"

Observations: Watching the Growth

Are the children able to recognize that different groups have the same number, even when they look different or are made of different materials?

Do they remember what number is hiding underneath, or do they choose randomly?

Sorting Attribute Blocks

Materials:

• Attribute blocks

Lesson Focus:

• To sort by various attributes

The Task:

Focus the children's attention on the attributes of color, shape, size, and thickness, and have them sort by these attributes.

Begin by having the children describe the blocks. Say: *"What can you tell me about this block?"*

"It's red."
"It's a triangle."
"It's smaller than the other ones."

Choose one of the attributes and find all the blocks that fit that category. For example, say: *"Let's find all the red ones."* Continue the activity by using one of the other attributes the children suggest.

Interactions: Deepening the Understanding

Start with obvious attributes in the beginning. Color is the easiest attribute for children to use for sorting. If any children are focused on only finding blocks that are identical, hold up two non-identical blocks and say: *"Are these blocks the same size? The same color? The same shape? What is the same about these blocks?"*

You may wish to challenge some children to see if they can sort objects that do not have a particular attribute. For example, you might say: *"Let's find all the blocks that are not yellow."*

Observations: Watching the Growth

Are the children able to sort objects that are the same color but are different shapes?

Do they identify more than one attribute when describing a block (for example, a yellow square)?

Can they see more than one way to sort a particular set of items?

When looking for blocks that are, for example, both red and a triangle, do they find all the red ones and all the triangles instead?

Developing Math Concepts in Pre-kindergarten

Sorting Collections with the Teacher

Materials:

- Collections of buttons, keys, pods, nuts and bolts, or other items

Lesson Focus:

- To sort by various attributes

The Task:

Have the children work together to sort collections. To help them identify related attributes, describe one attribute and ask: ***"Who can tell me something about one of the buttons?"***

"This one has bumps on it."

"Do all the buttons have bumps?"

"No, this one is smooth."

"Let's put all the bumpy ones here and all the smooth ones here."

After the children have finished sorting, have them sort another way. Push all the buttons back together and ask the same question: ***"Who can tell me something about one of the buttons?"***

Interactions: Deepening the Understanding

Sort by the categories that the children suggest. Continue to sort in many different ways.

Observations: Watching the Growth:

Do the children naturally see ways to sort, or do you have to make suggestions?

Are the children able to sort groups of objects, or are they finding pairs that match?

Do they sort with a purpose? Do they go beyond color and shape?

Are they flexible in determining categories?

Do they change categories in the middle of sorting?

Sorting Shapes

Materials:

- Discovery Blocks Tangrams Polygon shapes

Lesson Focus:

- To sort by various attributes

The Task:

Help the children become familiar with the attributes of different shapes, and work with them to sort them in a variety of ways. Then, have them describe why particular shapes belong together. Ask: ***"What do you notice about the shapes? Is that true for all the shapes? How else could you sort them?"***

The following are examples of some of the ways that children sort:

- Shapes that are the same
- Shapes that look like something: stairs, pizzas, and so forth
- Shapes that look pointy
- Shapes that have square corners
- Shapes with a piece missing

Interactions: Deepening the Understanding

While observing the children, you might ask: ***"Tell me about how you sorted the shapes. Why do these go together?"***

Observations: Watching the Growth

What do the children notice about the shapes?

What kind of language do they use to describe them? How precise are they in identifying the attributes?

Developing Math Concepts in Pre-kindergarten

Sorting Blocks

Materials:

- Building blocks or geoblocks

Lesson Focus:

- To identify and describe various geometric attributes
- To sort by various attributes

The Task:

Have the children examine different blocks. Focus their attention on the geometric attributes of the blocks and have them sort by those attributes.

To help the children focus on attributes, you could ask: ***"What do you notice about this block?"*** A child might respond, for example, by saying, "It's pointy."

Ask: ***"Are all the blocks pointy? Let's put the pointy ones here and the ones that are not pointy there."***

The following are other attributes the children might sort by:

- Size
- Sides (or faces) that are the same

Interactions: Deepening the Understanding

Continue to ask questions to get the children to look more closely at the geometric properties of the blocks. For example:

"What do you notice about this block? Are there any squares?"

"Does this block have a skinny rectangle or a fat rectangle? What about this block? Does it have any rectangles that look like this one?"

Observations: Watching the Growth

What do the children notice about the blocks?

Are they able to notice the different geometric features of the blocks, or do they sort only blocks that are identical?

Can they tell easily what shapes they see, or do they need to study the blocks before they can talk about what they noticed?

Sorting Boxes

Materials:

- Boxes of various sizes and shapes, with and without lids (small jewelry boxes, shoe boxes, cereal boxes, cracker boxes)

Lesson Focus:

- To sort by various attributes

The Task:

Help the children focus on the geometric attributes of boxes. Then, have them sort the boxes in various ways and find ways to describe their groupings.

To focus their attention on the geometric attributes of the boxes, begin by saying, for example: *"I see a square on this box. Do you see any other boxes that have square faces?"*

Or: *"I see that this box has a long rectangle. Do you see any other boxes that have a long rectangle?"*

Then ask: *"What did you notice about the boxes?"*

"I saw short ones."
"I saw skinny ones."

Interactions: Deepening the Understanding

Continue to ask questions to get the children to focus on the geometric properties of the boxes. For example, help children recognize the various shapes of the sides (faces). Accept the language individual children use if they are finding things that go together.

Some children will not be able to see one face or side as separate from the box as a whole. Using a pencil and paper, trace around some of the faces and let them see what happens. Let them see whether another box makes the same shape when you trace around it.

Developing Math Concepts in Pre-kindergarten

Observations: Watching the Growth

What do the children notice about the boxes?

Are the children able to see one part of the box while ignoring other shapes on other parts of the box?

Can they tell easily what the shape is, or do they need to study the box?

Are they able to notice geometric features, or do they focus on other attributes like size or color or any pictures on the box?

Do they need to touch the box in order to really "see" it, or can they look at it and describe what they see?

Copying Designs on a Geoboard

Materials:

- Geoboards
- Geobands

Lesson Focus:

- To copy the teacher's geoboard model

The Task:

Have the children copy geoband designs onto a geoboard to learn to be analytical about the various parts of a design and their position on the geoboard.

Build various shapes on a geoboard and ask the children to build them exactly like yours. Vary the designs' level of difficulty depending on the children's responses. (Designs involving diagonal lines are usually the most difficult for young children.)

Interactions: Deepening the Understanding

Ask: ***"Can you make a design just like mine?"*** Focus on the shapes rather than their size or position.

Observations: Watching the Growth:

Do the children produce approximate designs or exact designs, or are they unable to make a design that is similar to yours?

Grab Bag Geometry

Materials:

- Grocery bags or lunch sacks
- Various blocks and shapes, such as geoblocks, attribute blocks, polygon shapes, tangrams, or Discovery Blocks
- Common objects such as cans or balls

Lesson Focus:

- To explore the attributes of shapes by touching them without seeing them

The Task:

Have the children work with the objects you are planning to use for this task before you ask them to describe the objects when they are out of view. Discuss what they notice about the objects.

Help them find language to describe the attributes they are referring to. Ask them to do the following activity (identifying by touching) only after you feel they can talk about the objects when they are in view.

Activity 1: Find a Match by Touching

Have the children reach inside some bags and, without looking, try to find two objects that are the same. Then, have them take the two objects they have identified and put them on the table to see if they match.

Interactions: Deepening the Understanding

Help the children determine whether or not the items they have pulled out of the bag are the same. Sometimes laying one on top of the other will make it more obvious. If the objects are not exactly the same, help the children look for what is the same about them.

Observations: Watching the Growth

Are the children able to find matching pairs without looking, or do they take items out of the bag to check?

Do they find items that have some of the same characteristics but not all?

Activity 2: Guess What's in the Bag

You need two each of several shapes or objects. Put one set of objects in a bag. Have the matching objects lying outside the bag. Have a child reach inside the bag and describe one of the objects. Have the other children in the group try to find the matching object.

Interactions: Deepening the Understanding

Ask: **"Does it have corners? How many corners does it have? Is it round? Is it flat? Is it a square, a circle, a cube? How do you know?"**

Observations: Watching the Growth

What kind of language do the children use?

Can they find ways to describe the objects using their own informal language?

Do any of them tell what the object looks like rather than using the mathematical name? For example, do they say things like "It's like a door" or "It's like a witch's hat"?

Developing Math Concepts in Pre-kindergarten

Measuring Containers

Materials:

- Measuring Jars Station set-up: 2 jars labeled with different-colored dots, rice or other pourable material

Lesson Focus:

- To compare two jars to find out which holds more and which holds less

- To describe what they found out

The Task:

Present two jars to the children and ask: ***"Which of these two jars holds more?"***

Fill one jar to the top with rice and pour it into another jar. Ask the children, ***"Does this jar hold more than, less than, or the same as this jar? How do you know?"***

The children will be able to focus more on comparing containers if you pose the questions and comment on what is happening. A few children may be ready to work on this activity at the Measuring Jars Station. Some of these children will want only to find out what happens when they compare jars but will not be ready to focus on making a record of the experience. Other children will want to draw a picture of what happened.

Schedule ongoing opportunities for a few children at a time to work at the Measuring Jars Station, either with you or independently.

Interactions: Deepening the Understanding:

As the children gain experience with the jars, there will be times when the results surprise them. For example, they may be surprised when a short, wide jar holds more than one that is taller and skinnier. When the children get unexpected results, ask them what they think happened.

Observations: Watching the Growth:

How carefully are the children filling the jars and pouring the contents of one jar into another? Do they notice that their measurements are not accurate if they spill?

Do they comment on what happens?

Do they have any explanations for why one jar holds more or less than another?

When comparing jars, are they able to draw a picture that communicates their results?

***4:** Settings for Learning*

Comparing Objects Using Scales

Materials

- Scales
- Objects to weigh

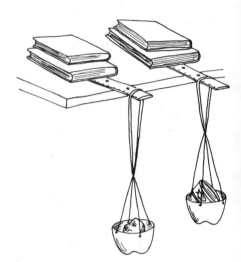

Lesson Focus:

- To compare the weight of two objects

The Task:

Show the children two objects and ask: ***"Which of these two objects is heavier? Which is lighter? How can we find out?"***

Have one child demonstrate by placing the objects in the scales. For example, a child might compare a rock and a can of tuna.

Most children will be able to focus on comparing the weight of two or more objects if the teacher is posing questions and commenting on what is happening. It is important for the children to learn to describe what happened orally: "The rock is heavier than the ball." It is more difficult for them to learn to describe the converse relationship: "The ball is lighter than the rock." As the teacher, you can model this language.

A few children may be ready to work on this activity at the Weighing Station. Some of these children will want to find out what happens when they compare the weight of two objects but will not be ready to focus on making a record of the experience. A few children may want to record what happened and will be able to do so independently.

Schedule many ongoing opportunities for a few children at a time to work at the Weighing Station with the teacher or independently.

Interactions: Deepening the Understanding

As the children are working, ask questions such as ***"Are the biggest objects always the heaviest? Did you find anything that is small and heavy?"***

Observations: Watching the Growth

Do the children assume big things always weigh more?

Do they comment on what they are discovering?

Do they make any predictions? What kinds of predictions do they make?

Is It Heavier or Lighter?

Materials:

- Homemade scales
- Objects to weigh

Lesson Focus:

- To compare many different objects to one particular object.

The Task:

Have the children choose something to weigh (for example, a can of tuna) and then have them weigh other things to determine whether they are heavier, lighter, or the same weight as the first object.

For example, have a child place the can of tuna in one tray of the homemade scales. Have another child pick an object to place in the other tray of the scales. Ask the children to tell what they noticed.

> "The tuna weighs more than the ball."

Then, keep the tuna in one of the scales and find a second object to compare. Ask the children to tell what they noticed.

> "The tuna is lighter than the rock." (Or "The rock is heavier than the tuna.")

Schedule many ongoing opportunities for a few children at a time to work at this task.

Interactions: Deepening the Understanding

Give the children opportunities to use comparative language by asking, "What did we find out?" You will notice that describing the comparisons can be difficult for children. If a child finds that something is heavier than the tuna, for example, they will be able to say, "The rock is heavier than the tuna." However, it will not be as easy for them to make the reverse comparison and say, "The tuna is lighter than the rock."

You may need to model the language for them. For example, you might say: ***You told me that the rock is heavier than the tuna. That means the tuna is lighter than the rock.***

Observations: Watching the Growth:

Are the children able to describe which object is heavier or lighter?

Do they comment on what they are finding out? Do they predict what is going to happen?

Setting 4: Individual Learning Time

Learning During Independent Station Time

Children should work at the individual learning tasks over and over again, so that they have the opportunity to get full value from these experiences. The children will stay interested and engaged as long as the tasks are appropriate for them and they are learning. After introducing the tasks to the children, have several tasks available, and allow them to choose which tasks they want to work on.

Do not begin the year with these activities. In the early part of the pre-kindergarten year, the most valuable thing for the children to be doing at station time is exploring materials. As the children become more and more familiar with activities such as patterning or independent number work, you will be able to provide choices from the following set of individual tasks for them to work on, with or without direct adult supervision, in addition to the exploration choices you provide. The tasks need to be available on an ongoing basis and should be repeated many times.

Individual Learning Time Activities:

Cover Them Up (Number), p. 138

Stories on the "Let's Pretend" Boards (Number), p. 140

Stories and Dice (Number), p. 142

Sorting Number Sets (Number), p. 144

Matching (Number), p. 146

Build One More (Number), p. 147

More or Less (Number), p. 148

Sorting More or Less (Number), p. 149

Pattern Station Task Cards (Pattern), p. 151

Sorting Collections Independently (Sorting), p. 154

Copy Cat (Geometry), p. 155

Matching Lids and Boxes (Geometry), p. 157

Recording Designs and Creations (Geometry), p. 158

Cut-and-Paste Geometry (Geometry), p. 159

Pattern Block Puzzles, p. 161

Cover Them Up

Materials:

Level 1: Counting:

- Counters
- Dot Cards (BLMs #4–13)

Level 2: Matching with Numerals:

- Counters
- Dot Cards (BLMs #4–13),
- Numeral Cards 1–6 and 0–10 (BLMs #14–17)
- Number Line Cards (BLM #18)

Lesson Focus:

- To focus on one-to-one correspondence
- To label groups with the appropriate numeral

The Task:

Level 1: Counting

As the children count, have them cover each dot on the Dot Cards with a counter. This task is particularly helpful for those children who need to focus on one-to-one correspondence.

If some children are unable to count independently, they can still work alongside other children covering dots. It would be helpful for these children to have an adult present to listen to them count or to count with them as they cover the dots. However, they can still practice matching dots and counters if no adult is available. This gives them an opportunity to feel that they can complete the job like the others are doing.

You can help ensure that these children complete the task by giving them more cards with 2 and 3 rather than cards with a higher number of dots.

Interactions: Deepening the Understanding

Say: *"How many on this card? Let's count together."*

Support individual children by helping them select appropriate cards, nudging them to try higher numbers or redirecting them to smaller numbers if necessary.

Observations: Watching the Growth

Do the child cover each dot once and only once?

Is the child covering the dots but unable to count up to particular numbers?

Is the child able to count increasingly higher numbers of dots?

The Task:

Level 2: Matching the Numerals

As the child counts, have the child cover each dot on the Dot Cards with a counter and then label each card with a numeral card to show how many she counted.

Interactions: Deepening the Understanding

Ask: *"How many did you count? Let's find the numeral that matches what you counted."*

Support individual children by helping them use the Number Line Card to recognize the Numeral Card they need.

Observations: Watching the Growth

Are individual children able to use the Number Line Cards or dots on the Numeral Cards to identify numerals they are unsure of? Or are they only able to use the Numeral Cards they already know?

Can they say the names of the numerals and build the appropriate sets when shown the numerals?

Stories on the "Let's Pretend" Boards

Materials:

- "Let's Pretend" counting boards: Colored paper to represent different environments (see page 28)
- Counters

Lesson Focus:

- To practice counting to a particular number over and over
- To use imagination to show how counting relates to what children know in their world

The Task:

Give each child a set of "Let's Pretend" boards (6 in a set) and counters. After individual children decide on what they want the counters to represent, have them put the same number of objects on each board.

The number to be worked with should be assigned by you according to the individual needs of the child. One child may need practice counting to 4, while another child may need practice counting to 9. You may want to give the child a number-set card showing the appropriate number of dots for his development.

This activity allows children to practice counting and using their imagination. If a child is working with the green "Let's Pretend" board, she may pretend that the counters are horses running in the field or children at the park. The child will be more imaginative and more involved in the "Let's Pretend" boards if you have discussed the possibilities during small-group work.

Developing Math Concepts in Pre-kindergarten

Interactions: Deepening the Understanding

Help the child focus on being accurate by saying:

"Can you count these for me?"

"Tell me your counting story."

"Can you show me how you counted?"

"Will you count that again?"

"Count that again and be as careful as you can."

"Let's count together."

When individual children accomplish the task quickly and easily and you realize they need a challenge, pose problems that require them to notice relationships. For example, if the child puts five counters on each board easily, ask her to try seven counters.

How does the child approach this task? Does she start over? Does she count each one and add on two or, after figuring out that two are needed, put two on each one without recounting?

What happens if you ask the child to make a smaller number—for example, use six counters instead of seven? This is a more difficult task, especially if the number the child is asked to make is more than five and thus not instantly recognized.

Observations: Watching the Growth

Watch to see how individual children accomplish the task. Is it hard to stay on task, or are they caught up in the task and fully attentive?

What size numbers are each of the children able to work with?

Are there any errors in the counting sequence?

Do they count each object once and only once, or do they lose the one-to-one correspondence?

Do they have a way of keeping track of what they have counted?

Are they accurate? Are they consistent? If inaccurate, are they bothered by that?

Do they check and recheck to make sure they counted right?

Stories and Dice

Materials:

- "Let's Pretend" boards: Colored paper to represent different environments (set of 6 for each child) (see page 28)
- Counters
- Level 1: Dotted dice
- Level 2: Numeral Cards 1–6 (BLMs #14–15)

Lesson Focus:

- To practice counting

The Task:

Level 1:

Have the child roll a die to determine the number of counters to be placed on each board. This is a more difficult task than counting out the same number of counters each time. The child has to be able to count or recognize the groups of dots on the die and switch from one number to another. You can vary the level of difficulty by changing the type of die (0–4 dots, 0–5 dots, 1–6 dots, 4–9 dots) provided.

There is no way for you to look at the child's completed work and know if the child has counted correctly or not. However, if the child is working at the appropriate level, he will be right most of the time. If you observe individual children actually working for a turn or two, you will get enough information to know if they are at the correct level.

Developing Math Concepts in Pre-kindergarten

Level 2:

Provide Numeral Cards 1–6 and have each child label his boards with the corresponding numeral.

Interactions: Deepening the Understanding:

Observe individual children at work to determine whether they are working at the appropriate level. If they are, they will be correct most of the time. Adjust the size of the dice, making the task harder or easier as necessary.

Help each child focus on being accurate. Say:

"Can you count these for me?"

"Tell me your counting story."

"Can you show me how you counted?"

"Will you count that again?"

"Count that again and be as careful as you can."

"Let's count together."

Observations: Watching the Growth

Watch to see how the child accomplishes the task. Is it hard to stay on task, or is the child caught up in the task and fully attentive?

What size numbers is the child able to work with?

Are there any errors in the counting sequence? Does the child count each object once and only once, or does she lose the one-to-one correspondence?

Do individual children have a way of keeping track of what they have counted? Are they accurate? Are they consistent? If inaccurate, are they bothered by that?

Do they check and recheck to make sure they counted right?

Sorting Number Sets

Materials:

- Number-set cards, portion cups, beads on a string, or other number-set materials (see pages 29–30)
- Two long pieces of butcher paper, each marked off into five sections labeled with dots 1–5 and 6–10

Lesson Focus:

- To count and then place number-set materials in the correct sections
- To notice how the same number can be represented in many different ways

The Task:

Have the child count using a variety of number-set materials and find the appropriate section on a large butcher-paper "number line." Several children can work on this task at the same time.

You can make the task simpler by using only the 1–5 butcher-paper number line.

Developing Math Concepts in Pre-kindergarten

Interactions: Deepening the Understanding

Help the child focus on being accurate. Say:

"Can you count these for me?"

"Can you show me how you counted?"

"Will you count that again?"

"Count that again and be as careful as you can."

Pointing to a section where the child has placed several cards, say: ***"Let's check and see if all these are fours, sixes, threes (and so on)."***

You can provide a challenge for those children who are ready for it by asking them to sort the number-set materials and put them in order without using the paper strip.

Observations: Watching the Growth

Children will approach the task in a variety of ways. Some will count each time, even when working with small numbers. They will search every time for the right place to put their number-set card without realizing that the big numbers are going to be farther down on the strip of paper. Some will need to recount the dots on the paper strip before placing their number-set card.

Some children will realize they can't count the big numbers and will be very careful to choose only those numbers they feel confident working with. Other children will sort the small numbers very quickly and easily but will have to be much more methodical when working with numbers above 6.

Watch to see how the child accomplishes the task. Is it hard to stay on task, or is the child caught up in the task and fully attentive?

What size numbers is each child able to work with?

Are they accurate? Are they consistent? If inaccurate, are they bothered by that?

Do they check and recheck to make sure they counted right?

Matching

Materials:

- Number-set cards (see pages 29–30)
- Collections of various items (buttons, keys, pods, nuts and bolts, and so forth)
- Several sheets of construction paper for each child

Lesson Focus:

- To practice making two sets of the same number of objects, either by using one-to-one correspondence or by counting

The Task:

Have the child choose one of the number-set cards. Then, have the child use the collections and construction paper to build a set that has the same number as the card. Repeat the activity several times.

Interactions: Deepening the Understanding

Say:

"Let's count and see if these are the same."

"I will point to one on this card, and you put one on the paper."

Observations: Watching the Growth

Watch to see how the child accomplishes the task. Is it hard to stay on task, or is the child caught up in the task and fully attentive?

What size numbers is the child able to work with?

Is the child able to copy the arrangement, or does he just count out a group of objects?

Build One More

Materials:

- Number-set cards or bead strings (see pages 29–30)
- Collections of various items (buttons, keys, pods, nuts and bolts, and so forth)
- Several sheets of construction paper for each child

Lesson Focus:

- To practice counting
- To practice making number sets that are one more than another set

The Task:

Have the child choose a number-set card or bead string. Then, have the child place a sheet of construction paper next to the number-set item and, using the collections, count out a set that has one more in it.

Interactions: Deepening the Understanding

Ask: ***"How many are on this card? Make this one first. Now what do you need to add? Now how many are there? Do you know how many you need to make one more?"***

Have the children label their cards with the appropriate numerals, if they are able.

Observations: Watching the Growth:

Some children will need to build a set that has the same as the number-set card or bead string, and then add one more and see what they have.

Others will be able to figure out the number they need ahead of time. For example, a child might say, "I have six. I need to get seven to make one more."

More or Less
(A game for partners)

Materials:

- Number-set cards (see pages 29–30)
- Construction paper
- More/Less spinner (BLM #19)

Lesson Focus:

- To compare groups of objects to determine which group has more and which has less

The Task:

Have each partner draw a card out of the pile of number-set cards and place their card on a sheet of construction paper.

One partner spins the More/Less spinner. If it lands on "More," the partner whose card has more wins and takes both cards. If it lands on "Less," the partner whose card has less on it wins the round and takes both cards.

The partners continue to play until they run out of cards.

Interactions: Deepening the Understanding

Ask: ***"How many are on this card? Is that more or less than your partner's?"***

If necessary, say: ***"Let's count together."***

Observations: Watching the Growth

Is the child able to tell immediately which is more or which is less, or does she need to count?

Is she accurate?

Does she need to check and recheck?

Sorting More or Less

Materials:

- Number-set materials (cards, beads on strings, or other) (see pages 29–30)
- 2 sheets of construction paper of different colors, labeled with "More" and "Less" symbols (for example, plus and minus signs) or More/Less Card (BLM #20) for each child

Lesson Focus:

- To determine whether groups of objects have more or less than the group they are being compared to

The Task:

Have the child lay the sheets of construction paper labeled "More" and "Less" side by side. Then, have the child choose one item from the number-set materials and place it between the two sheets of paper. The child then compares other number sets to the one chosen.

If the number-set is more than the chosen number-set, the child places it on the paper labeled "More"; if it is less, the child places it on the paper labeled "Less." If the number-set has the same number as the one originally chosen, the child places the item between the two sheets of paper, along with the original item.

Interactions: Deepening the Understanding

Help the children focus on being accurate. Say:

"What have you found out so far?"

"Can you show me how you counted?"

"Can you tell me which numbers are more than __?"

"Can you tell me which numbers are less than __?"

"Will you count that again?"

"Count that again, and be as careful as you can."

"Let's count together."

Observations: Watching the Growth

Watch to see how the child accomplishes the task. Is it hard to stay on task, or is the child caught up in the task and fully attentive?

What size numbers are individual children able to work with?

Are they able to use the language of "more" and "less" in describing how they sorted?

Do they need to count and compare each string before placing it, or can they place it without counting?

Do they have a way of keeping track of what they have counted? Are they accurate? Are they consistent? If inaccurate, are they bothered by that?

Do they check and recheck to make sure they counted right?

This task will be harder or easier depending on the types of materials and the size of the numbers used. If the child is sorting bead strings, he can use direct comparison and not have to count. If using random button cards or comparing toothpicks and buttons, the child will need to do some counting and more sophisticated comparing.

If you put all the cards out, you will see some children digging for the ones that they can handle and setting aside those that are too difficult.

Pattern Task Cards

Materials:

Using Pattern Blocks:

- Pattern blocks
- Pattern block pattern task cards*

Using Color Tiles:

- Ceramic tiles in different colors
- Color tile pattern task cards*

Using Unifix Cubes:

- Unifix cubes
- Unifix cubes task cards*

* See page 153 for directions and materials needed to make the task cards.

Lesson Focus:

- To copy and extend patterns

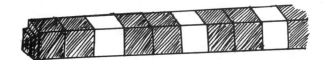

The Task:

Children need many opportunities to build and create patterns on their own. This activity assumes that the children have had opportunities to work with patterns in a variety of ways in teacher-directed lessons before tackling the work independently They also need to have had some experience interpreting the instructions for the individual activities (presented as task cards) before working on their own.

Encourage the children to make long patterns, because it is through repetition that they begin to get a true sense that patterns go on and on.

Posing the Problem:

Introduce the children to the pattern task cards and model copying and extending a pattern represented on one of the cards. Say: ***"Let's say the pattern on this card together."***

"Red, red, white; red, red, white."

Say: ***"Now tell me what I need to make the same pattern."***

"Red, red, white."

Ask: ***"Can I make the pattern longer? What comes next?"***

"Red."

"Next?"

"Red."

"Next?"

"White."

Interactions: Deepening the Understanding

If individual children are having difficulty copying a pattern, work with them to describe the pattern. Say: ***"Let's say the pattern together while I point to the blocks."*** (Example: "Red, red, green; red, red, green.")

Ask: ***"Will you show me how your pattern goes?"***

If necessary, help individual children find the appropriate level by saying: ***"Let's find another pattern card to use."***

Help each child understand that a pattern is predictable by saying: ***"Watch me make a pattern. What do you think is next? And next? And next?"***

Encourage the child to make long patterns by saying: ***"I want you to make your pattern very long. Can you make it go all the way from here to the table? I'll come back in a few minutes and see how long you have made it."***

Observations: Watching the Growth

What is the level of complexity of the patterns the child is working with?

Does the child simplify the pattern and always make an AB pattern no matter what pattern she is asked to make?

Does the child need to place objects directly on the card to copy the pattern, or can the child copy the pattern in the work space next to the card?

Can individual children extend patterns? Are they consistent when extending the pattern, or do they lose track as they make it longer?

Can they create their own patterns? What level of complexity do their patterns have?

Developing Math Concepts in Pre-kindergarten

Help them see that they can correct their own mistakes by saying: ***"Let's say your pattern together."***

"Red, red, white, black; red, red, white, black; red, red, white, ... Oops, I left out a black!"

Making the Pattern Task Cards

Pattern Block Pattern Task Cards

Materials:
- 3" × 9" tag board strips
- Colored paper to match the colors of the pattern blocks: yellow, orange, red, blue, tan, green
- Pattern Block Shapes (BLMs #37–42)

1. Copy the blackline masters onto the sheets of different-colored paper.
2. Cut out the pattern block shapes that match the color of the paper (for example, cut out the yellow hexagons from the yellow paper, red trapezoids from the red paper, and so on).
3. Arrange and glue down the shapes on the tag board strips to make patterns.
4. Make the patterns of varying levels of difficulty.

Color Tile Pattern Task Cards

Materials:
- 3" × 9" tag board strips
- Tiles made out of colored construction paper squares

1. Arrange and glue down the paper square tiles to make patterns.
2. Make the patterns of varying levels of difficulty.

Unifix Cubes Pattern Task Cards

Materials:
- 9" × 12" tag board
- Pattern Train Outlines (BLM #21)

1. Copy the blackline master and glue it to the tag board.
2. Color the Unifix cube trains to make patterns.
3. Cut the tag board apart to make 3" × 9" pattern task cards.

Sorting Collections Independently

Materials:

- Collections of various items such as buttons, keys, pods, nuts and bolts

Lesson Focus:

- To notice attributes and sort by them while working independently

The Task:

Ask: *"How can you sort these? Can you find the ones that go together?"*

Interactions: Deepening the Understanding

If the child hesitates and can't begin to sort on his own, ask: *"Can you tell me something about this button? What color is it? Can you find other buttons that are the same color?"*

Or you might ask: *"What did you notice about this key? Do all the keys have that?"*

When a child can make a set of identical items only (for example, two small blue buttons), ask: *"Do you have any other blue ones?"* (For example, there may be a larger blue button or another blue object in the collection.)

An important part of the task is to label the groups according to their different attributes. Ask individual children to describe how they are sorting. If they are unable to do so, verbally model this for them. For example, say: *"I see that these are the round buttons and these are the flower buttons. Can you tell me about these buttons?"*

You can help a child see a less obvious attribute by asking: *"Where would this one go? What about this one?"*

Observations: Watching the Growth

Does the child need to be directed to find categories of attributes, or can the child get started on her own?

Can the child identify attributes to sort by?

Developing Math Concepts in Pre-kindergarten

Copy Cat
(A partner activity)

Materials:

- Pattern blocks, geoblocks, geoboards, tiles, wooden cubes, Discovery Blocks, tangrams, or other building blocks

Lesson Focus:

- To create and copy structures or designs, working with a partner

The Task:

As the children work with materials on their own, some will naturally invent "Copy Cat" activities. Children will often copy their own designs as well as the designs of other children when they find it a challenge to reproduce a structure.

If none of the children begins reproducing structures and designs on his own, you can suggest or model the task yourself.

One way to introduce the idea is to sit down beside individual children as they work and ask them to build a design that you can copy. After you have copied the child's structure, you can then build a design for the child to copy. Other children will see this and begin to work in the same way.

···

Being able to copy a design requires children to see the parts of the design and their relationships to other parts. Children become intrigued with this kind of activity when it poses the right level of difficulty for them. Say: ***"See what I made? Can you make another one just like it?"***

Interactions: Deepening the Understanding

Because the children are creating the structures, there is no limit on how simple or how complex their structures can be. Most of the time, children naturally find the level that works for them and their partner. However, if you see individual children building very simple structures that do not challenge their partner, you may suggest that they add on to their structure by saying: ***"Can you make your design bigger?"***

Some children will find the structures their partners make very difficult to copy. When this occurs, ask the child's partner to add just one block at a time so that the child can reproduce what they are building.

Observations: Watching the Growth

Do the children have to stop and look again to see how to place a block?

Do they have to try more than one way?

Do they notice if they made a mistake?

Do they actually have to move to get a better view of the design?

Matching Lids and Boxes

Materials:

- Boxes of various sizes and shapes, with and without lids (Examples: small jewelry boxes, shoe boxes, cereal boxes, cracker boxes)

Lesson Focus:

- To notice size and shape in matching boxes and lids

The Task:

Help individual children notice size and shape by asking them to match the boxes and their lids. Ask: ***"Can you find the lids to match the boxes?"***

Interactions: Deepening the Understanding

While individual children are work-
ing, discuss with them how they know which boxes and lids go together. Have them think about whether they had to try matching the lid to the box first or if they could tell which lid matched which box just by looking. Ask: ***"Are some boxes and lids easier to match than others?"***

Observations: Watching the Growth

Do individual children see easily which lid matches which box, or do they need to try them out first?

Do they notice differences in size before trying to match the lids to the boxes, or do they try to make them fit if they are a similar shape?

Recording Designs and Creations

Materials:

- Tiles or pattern blocks
- Paper shapes that match the tiles or blocks
- Paper and glue

Lesson Focus:

- To learn to consider shape and position by building designs and then copying them with matching paper shapes

The Task:

Have the child create designs with the blocks. Then, have the child copy these designs by gluing down paper shapes that match the blocks. It is important that children actually build with the blocks first and not simply create a design using the paper shapes. The challenge in this task is to reproduce the design. This requires the child to pay close attention to the position of the blocks relative to each other and the number being used.

Posing the problem:

Ask: **"Can you make a copy of your creation using these paper shapes?"**

Interactions: Deepening the Understanding

Often children will create three-dimensional designs that are too complex for them to reproduce using paper shapes. To help limit the size of their designs, give them a 4" x 6" sheet of paper on which to build with the blocks. You may also limit the number of blocks they can use.

Observations: Watching the Growth

Do individual children have to stop and recheck to see how the paper shapes match their block design?

Do they have to try more than one way?

Do they notice when they make a mistake? How accurate are they?

Cut-and-Paste Geometry

Materials:

- Various jar lids
- Large attribute blocks or other blocks
- Boxes (used to create shapes by tracing around the sides)
- Construction paper, paste or glue, pencils, scissors

Lesson Focus:

- To explore shapes by creating figures and designs with the shapes

The Task:

In this activity, the children will have experiences that integrate art and geometry by creating figures and designs using paper shapes such as rectangles (including squares), circles, and triangles of various sizes.

Have the children cut the shapes apart and create new shapes. Ask:

"What can you make with these shapes?"

"What can you make using only circles? Only squares?"

"What new creations can you make when you cut the circles or other shapes apart?"

Version 1: Tracing Lids and Blocks to Make Shapes

Have the children trace around lids to make circles and around blocks and boxes to create the other shapes they want to make.

Children are often very creative when given the opportunity to combine shapes into figures and structures of various kinds. Their ability to do this will increase over time if you just provide the opportunity.

Interactions: Deepening the Understanding

Ask: *"What blocks (lids, boxes) did you use to make your picture?"*

Say: *"Tell me about what you are making."*

Observations: Watching the Growth:

How does the child proceed with the task? Does the child start building with little or no obvious planning? Does he randomly place the pieces with little or no attention to symmetry?

Do individual children build first and then decide what it is they made? Or do they start with an idea and build toward it?

Do they seem to know what they need to do to follow through on their idea?

Do they add details to their creations? Do they find unusual ways to create these details?

Are the relative sizes of their creations appropriate or not?

Version 2: Copying Models

After the children have had opportunities to work with the lids and blocks in their own way, provide models of things that the children can copy or adapt with the shapes they have made, such as:

Developing Math Concepts in Pre-kindergarten

Pattern Block Puzzles

Materials:

- Pattern blocks
- Pattern Block Puzzle Cards (BLMs #22–36)

Lesson Focus:

- To explore how shapes can be made by other shapes
- To find ways to fit shapes into puzzle outlines

The Task:

Level 1: Work with the Level 1 Pattern Block Puzzle Cards that show the outlines of the blocks that fit, as shown below. Have the child fit pattern blocks into the outlines shown.

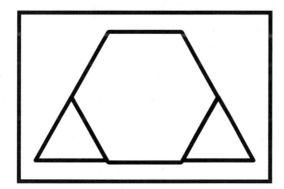

Level 2: Work with the Level 2 Pattern Block Puzzle Cards that accommodate various arrangements and types of blocks, as shown below. Have the child find blocks that he or she thinks will fit and fill in the shapes.

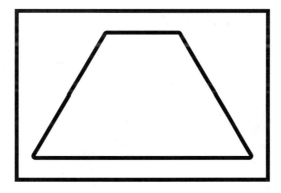

Interactions: Deepening the Understanding

Level 1:

Can you find a block that will fit in this space?

What block will fit in this space? Let's try it and see.

Can you turn the block to make it fit?

Level 2:

Can you move this block so that another one can fit, too?

Observations: Watching the Growth

Do the children fit the blocks into the puzzle, or do they just lay the blocks inside the puzzle without worrying about whether they fit or not?

Do they pay attention to the boundaries, or do they let the blocks go over the lines?

Do the children test the blocks and move them around to make them fit?

Do they know ahead of time what blocks to get, or do they just try out blocks to see what happens?

When working with the Level 2 puzzles, do they actually put blocks together to make them fit inside the outline, or do they just lay blocks down that can fit inside the shape?

Setting 5: Learning with Others Throughout the Day

Much of the learning that takes place in the pre-kindergarten classroom will happen as situations arise. Take advantage of those moments and help children to see how mathematics is a part of much of what they do all day long.

There will be many opportunities during the natural course of the day to count and to solve problems. There will be other times that you can set up situations that lend themselves to using mathematics, such as growing plants or preparing and sharing snacks.

Throughout-the-Day Activities:

Counting, Counting, Counting

Materials:

- Objects or people in the child's environment

Lesson Focus:

- To practice the counting sequence and extend it beyond the numbers already memorized

The Task:

Lead the children in counting experiences as opportunities arise throughout the day. Frequent whole-class counting experiences provide appropriate practice for children who have already memorized the counting sequence to ten and those who do not yet count to ten independently. All children need to see the sequence they know in the context of the longer counting sequence.

Opportunities to hear this longer sequence will arise in the natural course of the day in the following types of activities:

Counting the children

Take advantage of the many opportunities that arise to count the children for a variety of purposes. Ask, for example:

"How many are here today?"

Developing Math Concepts in Pre-kindergarten

"How many boys? How many girls?"

"How many are buying lunch? Milk?"

"How many brought back their notes?"

Counting things in the room

"How many books are we going to return to the library today?"

"How many crackers did Jimmy bring for snack today?"

"How many did Leila bring?"

Counting collections

"How many leaves did we collect on our walk?"

"How many shells did Joey bring for the science table?"

Interactions: Deepening the Understanding

Be sure to organize the counting so that it models one-to-one correspondence. For example, make it obvious which child you are counting by pointing to the child as the group counts along with you.

You can involve the children in this by having them indicate when they have been counted. For example, they might have their hands up and then put their hands down when they have been counted. Or, they could all be standing and then sit down or move to a different place when they have been counted.

Focus the children on the need for accuracy by asking questions when they stray from one-to-one counting. For example, if two children raise their hands at the same time or move to the new spot, say: ***"Did we count everybody one time? Let's check and see. How many do you think we will have if we count again?"***

Vary the pace of the counting so that the children have to pay attention.

Observations: Watching the Growth

Are the children able to count in unison, or do some rush ahead or drop out of the counting?

Are they bothered if they end up with a different number when recounting the same group of objects?

Just Enough

Materials:

- Classroom supplies and/or snacks

Lesson Focus:

- To figure out ways to get just enough

The Task:

Take advantage of opportunities that arise during the day to ask groups of children to problem-solve getting just enough of whatever is needed for their group. For example, say: ***"Everybody needs a straw for their milk. How can we figure out how to get just enough?"***

This matching of sets can help children move toward understanding quantity. As their thinking becomes more refined, they begin to notice when they have too many or too few, when one group is more or less than another, and even how many more or less one group is than another.

This kind of matching of objects is most powerful when it happens in the context of real needs. Don't belabor it if the real purpose of a lesson does not include matching, but if you are going to point it out, include it in your mind as part of the math lesson and give it its due.

It is important to have all the children in the group involved in the thinking and figuring out. Do not make it into a task that you give one child to do while the rest wait. For example, if Bobbie is going to be the one to get the chalkboards, involve everyone in figuring out how many she needs.

Interactions: Deepening the Understanding

Get the children's ideas on ways you can make sure you have just enough. Depending on what their ideas are, pose questions such as *"Can we find out without counting?"* or *"Can we use counting to help us?"*

Vary the task by saying: *"I have six pieces of chalk (or paper, bowls, straws, and so on). I want everyone in the group to have just one. Do you think I have just enough, too little, or too much?"*

Observations: Watching the Growth

This type of problem can be solved at many different levels. Some children will find and deliver one item at a time until the job is done. Other children will count the group and get the number needed.

The task provides an opportunity for those children who are not confident counters to use what they do know and understand. Even before children can determine exactly "how many," they begin to get a notion of number by matching sets of objects.

I See a Pattern!

Materials:

- Objects or people in the child's environment

Lesson Focus:

- Recognizing and describing patterns as situations arise

The Task:

You can tune children in to the many patterns in their world by encouraging them to tell you when they see a pattern. The whole class can then describe the pattern together if it is an appropriate time to do that.

The children will see patterns on people's clothing, on books, and in the classroom itself. For example:

"Look! B.J. has a pattern on her shirt."
"This book has diamond, square, diamond, square on the front."
"The floor has a pattern."
"The windows have a pattern."

Interactions: Deepening the Understanding

Call attention to patterns when providing directions in many different settings, such as:

- Lining up by boy, girl, boy, girl
- Decorating the border on place mats made for special occasions when visitors come to school
- Arranging snacks in a pattern

Help the children describe and extend the patterns. Say: ***"Let's describe the pattern together. How does it start? What would it be if it was longer?"***

Observations: Watching the Growth

Is the child able to see more and more patterns over time?

Can the child distinguish between predictable patterns and designs?

Let's Measure It!

Materials:

- Objects or people in the child's environment

Lesson Focus:

- Measuring and comparing as situations arise

The Task:

Have the children compare time and distance informally as the situations arise. Say:

"Do you think it is farther to the library or to the playground?"

"Do you think it takes longer to walk to the bus stop or to the cafeteria?"

"Let's see how long it takes us to get our coats on and get ready to go. The long hand is pointing at the 5. Let's see where it is pointing when we get done."

"I think this pumpkin weighs more than that pumpkin."

"We poured eight glasses of lemonade from this pitcher."

Encourage the children to measure and compare when playing with the blocks or other math materials. For example, they might say:

"Look. I'm taller than my building."

"This Unifix cube train is longer than Joey and me."

Measure as part of science activities. For example:

- Measure and keep track of how much the classroom plants have grown.
- Fill a jar with water and mark the water level each day as it evaporates.
- Weigh the class pet to keep track of its growth.

Appendix A:
Planning Guide for Pre-Kindergarten Math Time

Planning Guide for Pre-Kindergarten Math Time

While we know that mathematics occurs everywhere in the world around us, it is still necessary for children to have a particular time in the school day when they can work with mathematical ideas. We can ensure they get the experiences they need if we establish a set math time during which the whole class is focused on mathematics.

Math Time will usually consist of the following blocks of time:

Circle Time (approximately 10 minutes)

Exploration Time and/or Individual Learning Time (20–30 minutes)

Teacher-Directed Small-Group Learning Time (5–10 minutes during Individual Learning Time or at another time of day)

Mathematics learning will also occur as situations arise throughout the day.

Math Time begins with whole-class Circle Time and then moves on to independent work as the children explore math materials using their own ideas (Exploration Time) or work on specific tasks (Individual Learning Time). There will also be times when you work with small groups of children on focused tasks that meet specific needs.

When planning Math Time, it is important to remember that children deepen their understanding of math concepts by repeating many of the same core activities throughout the year. They don't always need new tasks in order to stay engaged and to learn. They benefit from exploring new questions and developing new understandings using the same activities. We can meet children's needs over time by adjusting the size of the numbers, the complexity of the patterns, and the difficulty of the tasks.

The experiences presented in this book are meant to be done over and over again. Over time the children will develop deeper insights and higher levels of competency. Included with each activity are suggestions for interactions with the children that will help deepen their understanding and a guide to what to look for to determine each child's growth.

This Planning Guide provides a structure that will help you make sure you are including the variety of experiences needed to benefit your students. It is important to note that, as the classroom teacher, the instructional decisions you make on a daily basis can only be made based on the children's needs and their developing interests. You will need to increase the size of the numbers being learned and the complexity of the tasks according to what you see your students doing. Most tasks lend themselves to multiple levels of difficulty. You may find that, for some children, you need to continue to offer some activities at an easier level, while beginning to offer more complex versions of the same activities as other children become ready.

I. Beginning the Year

Circle Time

Circle Time helps you establish the comfort of a routine by doing certain activities every day. It is also a time to expand the children's thinking by providing activities that may be familiar but are not done every day.

This is the time to gather the children together to begin the day and to:

- Make note of the date,
- Become familiar with the repeating cycle of the days of the week,
- Notice the passage of time by keeping track of the number of days in school, and
- Make a daily record of the weather.

It is also the time to provide ongoing practice in learning the counting sequence, to develop an awareness of pattern, and to begin to look for likenesses and differences in objects throughout the classroom.

Daily Routines:

Take the time to do these activities every day:

What Day Is It Today?, p. 66
How Many Days in School?, p. 69
Weather Graph, p. 73

Feel the Beat, p. 76
Take a minute or two every day to practice the counting sequence. Focus on the needs of the children who do not yet know the sequence. You may need to start with counting to 3 or 4.

Rhythmic Patterns, p. 70
Introduce the concept of pattern through whole-class experiences with rhythmic patterns. Include a variety of patterns. Do not overemphasize AB patterns. Make sure to include patterns like AAB or ABC patterns. Do not worry that only a few children are able to do the patterns at first. Over time, more and more children will be able to join in. It is easier for young children to follow a pattern if you name the part of the body they are to touch as they do the pattern. For example, say, "Nose, nose, chin, nose, nose, chin."

Other Activities to Use:

Do one or both of the following activities each day:

Counting Poems and Finger Plays, p. 74
Counting Jars, p. 78

Exploration Time

Children's first independent learning experiences will be those of exploring what they can do with the math materials using their own ideas. The informal exploration of geometry, sorting, number, and ordering is the heart of young children's work.

During this critical exploration time, children learn the attributes of the various materials they are exploring and naturally find those that go together. They work with plane and solid geometric shapes and learn how shapes fit together and what they can do with them. They are also becoming familiar with the Unifix cubes, colored tiles, toothpicks, and number set materials you will be using later for structured tasks.

Use Exploration Time to establish the procedures you want the children to follow whenever they work independently. What you do now to establish how the children should work while exploring materials will help you later in the year when you provide more structured tasks for them to work on independently.

Many of your interactions with the children will be geared to help them develop the social behaviors necessary to work productively with the materials. They will need to learn how to share, how to stay engaged, and how to clean up when they are finished with a material.

You will also be interacting with the children in ways that will help them focus on the mathematics in their work. See each activity for suggestions for questions you can ask and observations you can make.

Introduce the materials to the whole class.

Introduce each material to the whole class before making it available for children to work with independently. Talk about what the material is called and where it is stored. Solicit ideas from the children and discuss the kinds of things they might be doing with the particular material. Ask questions to get them thinking about the potential of the material, such as "I wonder if you could make a tower with these? Or fit them together to make a road? Or fit one box inside another box?"

When you introduce the materials by asking questions and wondering aloud what can be done with them, you model the thought process you want the children to use when doing the task on their own. The goal is for the children to continue asking questions themselves, even when working independently.

Exploration Time

Activities to use:

Exploring Building Blocks, p. 48

Exploring Boxes, p. 51

Exploring Small Blocks, p. 52

Exploring Collections, p. 56

Exploring Counting Materials: Unifix Cubes, p. 53

Exploring Counting Materials: Colored Tiles, p. 53

Establish expectations:

Teach the children the following procedures and expectations. These same procedures and expectations will apply when they begin working on the structured activities you make available during Individual Learning Time.

Have the children set up.

If you have the materials stored in tubs, you can have the children set up the materials at particular places in the room. If you label the tubs with symbols such as circles or triangles, you can post matching labels in the areas of the room where you want the materials to be delivered. Some materials will work best on the floor and some will work well on tables.

Let the children choose where they want to work, and let them move from place to place whenever they are ready to work with something else.

One of the goals of Exploration Time is to have children learn to stay involved and engaged with the task at hand. Children will be more likely to have a high level of involvement if they have made the selection of what they want to work with. Sometimes they need to try out a couple of places to work before they can get settled in. You just need to give them time to figure this out.

However, if a child persistently cannot make a choice and tends to wander, you can offer a narrower range of choices. You might say something like "Do you want to use wooden cubes, or do you want to work with the pattern blocks?" If they still have trouble getting engaged, offer some suggestions. You can also set a goal for them and come back to see if they have accomplished the goal. For example, you could say, "I would like to see if you can make a road all the way to the edge of the rug. I will come back later and see how you did."

It is more important that children be engaged in what they are doing than that they experience every material available. Over time, most of them will move around and make new choices.

Let the children come back to the same material as many times as they wish to.

Sometimes, if children "stuck" and always goes to one or two places, you might want to invite them to work with you with a new material to see if you can help them see the potential of different materials.

Let the children work alone or with others.

Sometimes children will naturally work together with other children who have chosen the same materials, but other times the children will want to pursue their own ideas. Both ways of working are valuable to the child.

Reassure the children that they don't need to come and get you to show you what they are working on. The children will benefit from your involvement and interest in what they are doing. However, you also want to foster independence so you will be free to work with individual children and small groups when appropriate. Let them know that you will be coming around to see what they are doing, so they don't need to worry about that. Then make sure you move around the room, and let them see that you are interested in their work. Over time, they will become more and more focused on what they are doing without needing to show it to anyone or get approval from anyone.

If a problem arises, encourage the children to talk to each other to see if they can solve it before bringing it to you. You can role-play ways for them to figure out what to do if two people want the same block or intrude on each other's space.

If there are persistent problems that children need help solving, you will need to deal with them during whole-class time so that all the children know how to cope with the particular situation.

Intervene if necessary.

Sometimes children say they are bored when they mean they want some adult attention. They may not be used to taking responsibility for themselves. They could be having trouble making a choice, or perhaps some are having difficulty accepting the fact that they didn't get their first choice. Some children just need to be redirected. You could say something like "I would like you to choose a task, and I will come by later and see how you are doing." Once in a while you may find it helpful to work with individual children for a few minutes on a task they choose.

Have the materials available for the children to work with for several weeks.

It is sometimes hard to allow the children the time they need to learn all they can from working with the materials in their own way. But if the children are going to truly experience the potential of the materials, the same materials should be available for them to work with for several weeks at a time.

Small-Group Time

At the beginning of the year, introduce the children to small-group work through sorting activities that help them develop the language that describes the attribute blocks:

Color: *Red, blue, yellow*
Shape: *Triangle, square, rectangle, circle*
Size: *Small, large*
Thickness: *Thick, thin*

Activity to Use:

Sorting Attribute Blocks, p. 124

II. After a Few Weeks Have Passed

Exploration Time / Individual Learning Time

Continue:	Add:
Exploring Building Blocks, p. 48 **Exploring Boxes,** p. 51 **Exploring Small Blocks,** p. 52 **Exploring Counting Materials: Unifix Cubes,** p. 53 **Exploring Counting Materials: Color Tiles,** p. 53 **Exploring Collections,** p. 56	Over several days, present additional materials for the children to explore, using these activities: **Exploring Counting Materials: Wooden Cubes,** p. 53 **Exploring Counting Materials: Toothpicks,** p. 53 **Exploring with Mirrors,** p. 54 **Exploring Geoboards,** p. 58 **Exploring at the Measuring Jars Station,** p. 59 **Exploring at the Weighing Station,** p. 61 *Individual Learning Time:* Begin including some independent tasks during Exploration Time by setting up the tasks at learning stations around the room. Introduce these new activities over a period of several days. **Copy Cat,** p. 155 Make copying one another's designs a part of Exploration Time. **Matching Lids and Boxes,** p. 157 Set up one station with boxes with lids and have the children find the matching boxes and lids.

Small-Group Time

After a few weeks have passed, you will find that you have identified particular needs and will want to form small groups to meet those needs. Working with a small group allows you to really watch, interact with, and respond to individual children during a "teaching" time. Small-group instructional time is particularly important for children who find it more difficult to be engaged and focused when they are part of a large group.

You can use this instructional time in several ways: to meet with a small group that needs extra help, to meet with a group that is ready for more challenges than the rest of the children, to provide more focused practice for children than they can get working with the large group, or to introduce the children to something new that they will be doing independently later.

Try to be flexible in the use of small-group time. Groups do not need to be set up permanently but can be formed as particular needs arise. You can meet for very short periods with some groups and for longer periods with others. You may want to meet with some groups for several days in a row if they need extra help or an extra challenge.

Activities to Use: Sorting:

Sorting Collections with the Teacher, p, 125

Sorting Shapes, p. 126

Activities to Use: Counting:

Gather children with similar needs together and work with them using two or three of the following activities during a short small-group lesson. Choose which numbers to work with according to the needs of the children. Repeat these lessons for several days, choosing new activities to provide a variety of experiences.

Ding!, p. 100

Dump It Out, p. 101

Shake Them Up, p. 102

Making Buildings, p. 103

 "Let's Pretend" Stories (Level 1), p. 105

Acting Out Songs and Poems, p. 106

Activities to Use: Pattern:

Pattern Task Cards, p. 151

Work with the Pattern Task Cards activity during small-group time with any children who are ready for it, before including it as an independent learning-station activity.

Individual Learning Time

Continue:	Add:
Copy Cat, p. 155 **Matching Lids and Boxes,** p. 157	**Cover Them Up,** p. 138 **Stories on the "Let's Pretend" Boards,** p. 140 **Matching,** p. 146

Throughout the Day

	Add:
	Counting, Counting, Counting, p. 164 **Just Enough,** p. 166 **I See a Pattern!,** p. 168 **Let's Measure It,** p. 169

III. Moving On

Circle Time	
Continue:	**Add:**
Daily Routines:	**Descriptions,** p. 82
What Day Is It Today?, p. 66	**Real Graphs/Name Graphs: Name Graphs,** p. 89
How Many Days in School?, p. 69	**What's the Same as Me?,** p. 93
Rhythmic Patterns, p. 70	**Is It Longer or Shorter?,** p. 95
Weather Graph, p. 73	
Feel the Beat, p. 76	
Continue to present the concept of pattern through whole-class experiences. You will be able to present more complex patterns than earlier in the year, as more children will be able to join in.	
Other Activities to Choose From:	
Counting Poems and Finger Plays, p. 74	
Counting Jars, p. 78	
Alike and Different, p. 80	
People Patterns p. 84	
Real Graphs/Name Graphs: Real Graphs, p. 88	
Geometry Walks, p. 91	

Exploration Time

Continue:	Add:
Exploring Building Blocks, p. 48	
Exploring Boxes, p. 51	
Exploring Small Blocks, p. 52	
Exploring Counting Materials: Unifix Cubes, p. 53	
Exploring Counting Materials: Wooden Cubes, p. 53	
Exploring Counting Materials: Toothpicks, p. 53	
Exploring Counting Materials: Color Tiles, p. 53	
Exploring with Mirrors, p. 54	
Exploring Collections, p. 56	
Exploring Geoboards, p. 58	
Exploring at the Measuring Jars Station, p. 59	
Exploring at the Weighing Station, p. 61	

Small-Group Time

Continue:	Add:
Activities to Use: Counting	Some activities can be done more effectively in small groups than as a whole class. Provide the children with small-group learning opportunities through these activities:
Continue to work on counting tasks, changing the size of the numbers as children progress.	
Ding!, p. 100	**What's Hiding?,** p. 118
Dump It Out, p. 101	**How Many Do You See? Using Dot Cards,** p. 120
Making Buildings, p. 103	**How Many Do You See? Using Number Sets,** p. 121
Acting Out Songs and Poems, p. 106	**Can You Find It?,** p. 122
One More, p. 114	**Sorting Collections with the Teacher,** p. 125
Fingers and Cubes, p. 116	**Copying Designs on a Geoboard,** p. 130
Taking Turns, p. 117	**Measuring Containers,** p. 133
	Comparing Objects Using Scales, p. 134
	Is It Heavier or Lighter?, p. 135
	Focus on numeral recognition using the following tasks:
	"Let's Pretend" Stories, p. 105
	I Changed My Mind, p. 110
	Roll Again!, p. 112

Individual Learning Time

Continue:	Add:
Cover Them Up, p. 138 **Stories on the "Let's Pretend" Boards,** p. 140 **Matching,** p. 146 **Pattern Task Cards,** p. 151	Work on the counting activities with small groups before including them as independent tasks at the learning stations. After children are very familiar with the tasks, include them as choices during Individual Learning Time. **Stories and Dice,** p. 142 **Sorting Number Sets,** p. 144 **Matching,** p. 146 **More or Less,** p. 148 **Sorting More or Less,** p. 149 **Sorting Collections Independently,** p. 154 **Recording Designs and Creations,** p. 158

Throughout the Day

Continue:	Add:
Counting, Counting, Counting, p. 164 **Just Enough,** p. 166 **I See a Pattern!,** p. 168 **Let's Measure It!,** p. 169	

IV. The Final Weeks of the Year

Choose favorite activities as well as provide activities that children were unable to do earlier in the year.

Circle Time

Measuring with Strings, p. 97

Small-Group Time

Concentration, p. 123

Copying Designs on a Geoboard, p. 130

Grab Bag Geometry, p. 131

Individual Learning Time

Build One More, p. 147

More or Less, p. 148

Sorting More or Less, p. 149

Cut-and-Paste Geometry, p. 159

Pattern Block Puzzles, p. 161

Appendix B: Blackline Masters

Ten Frames

Ten Strips   BLM 3

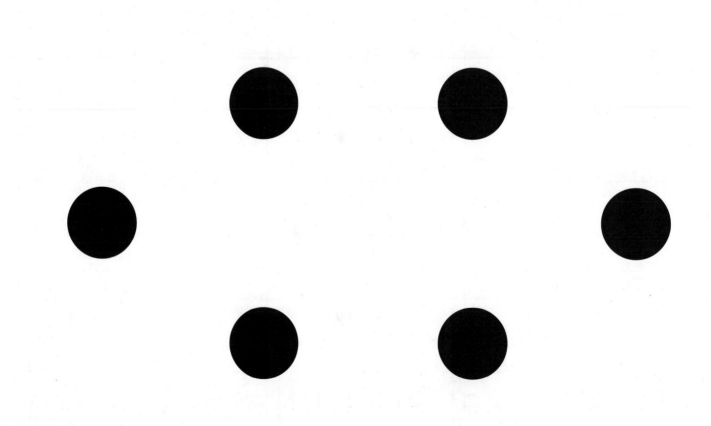

Dot Cards

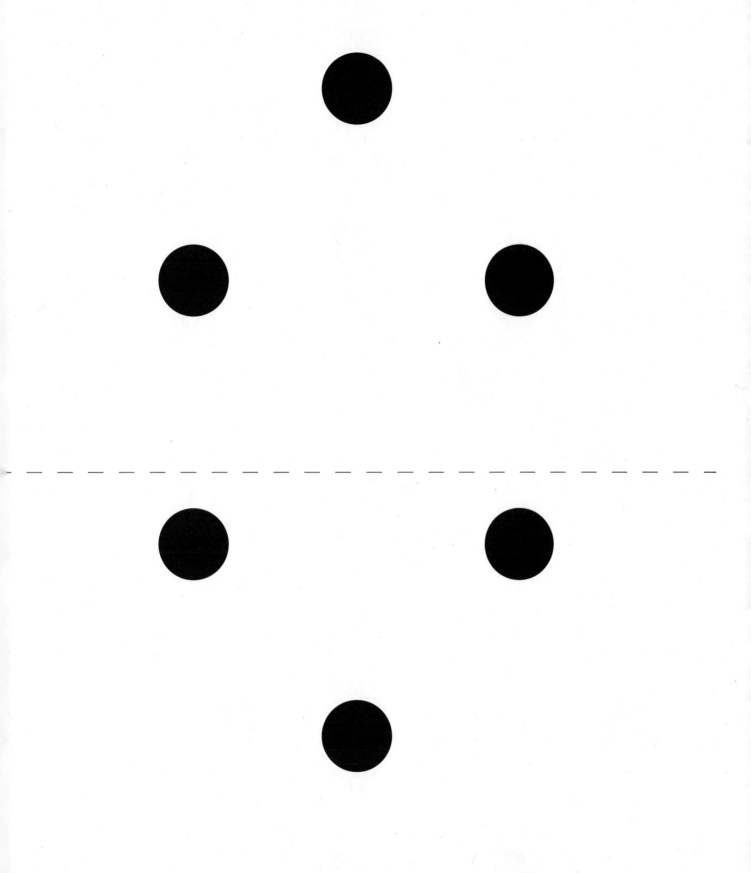

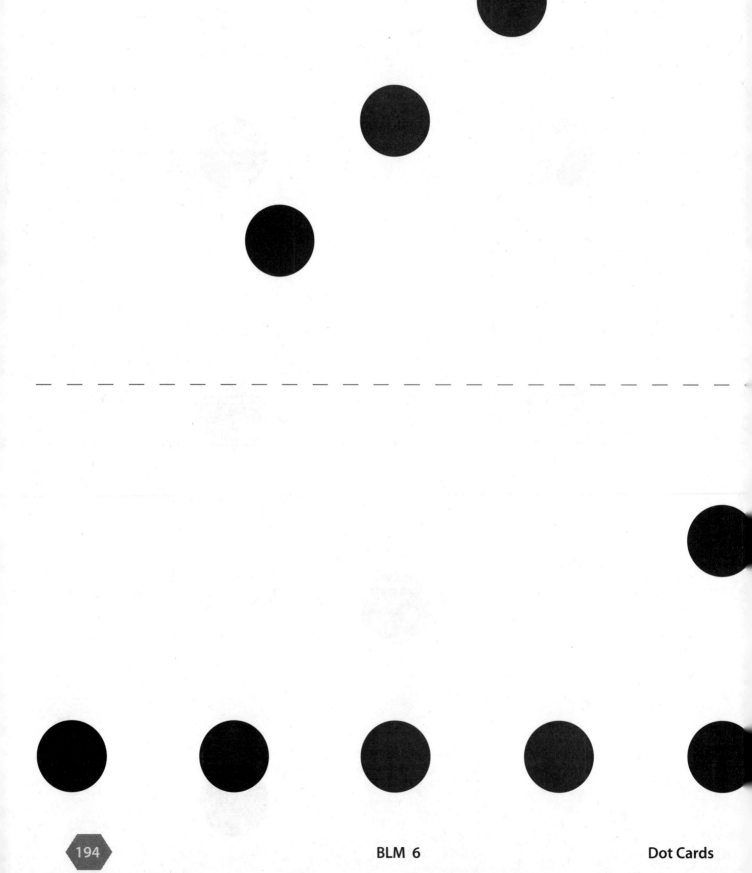

BLM 6

Dot Cards

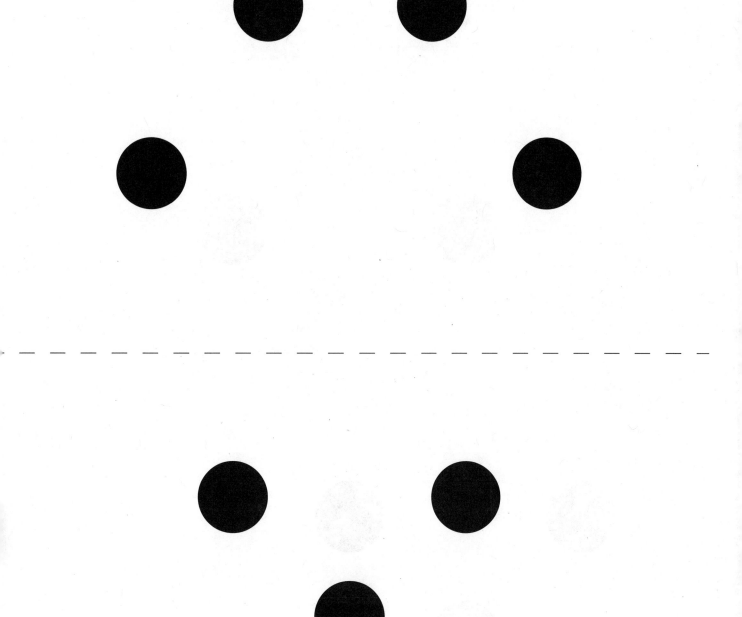

BLM 8

Dot Cards

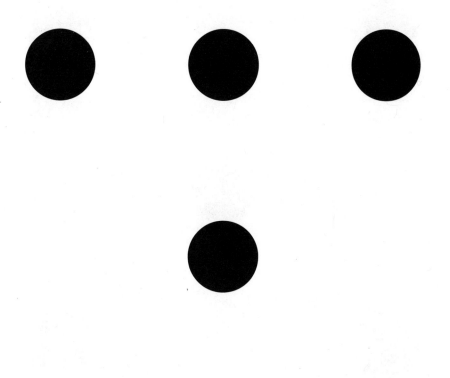

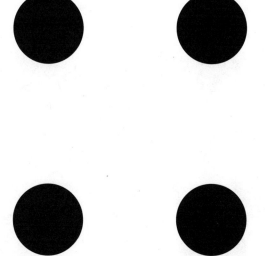

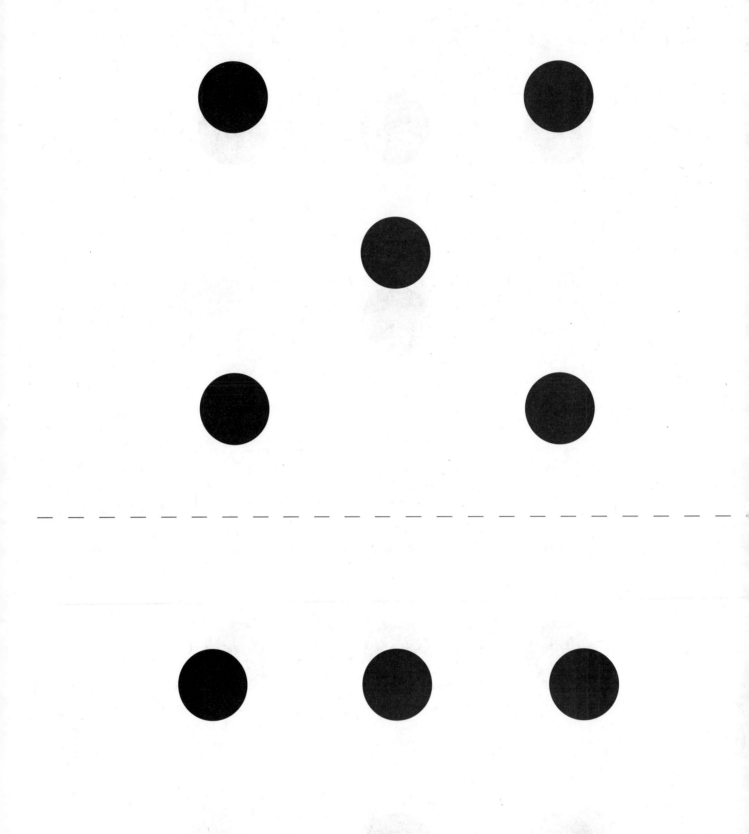

Dot Cards

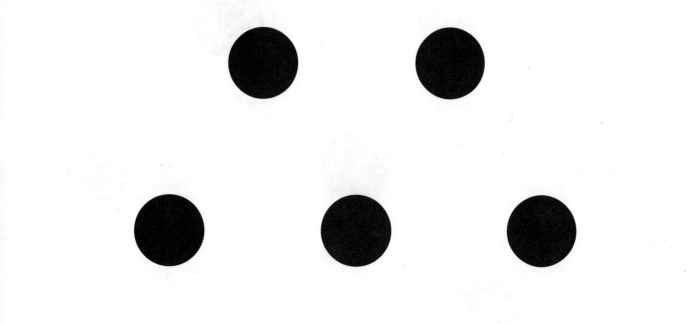

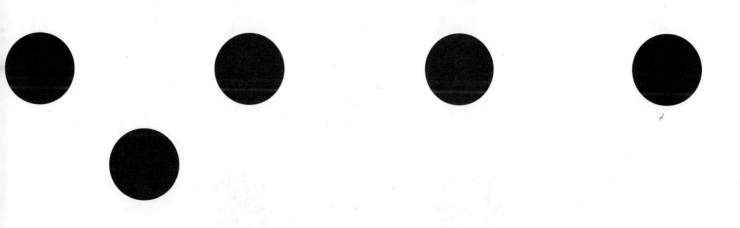

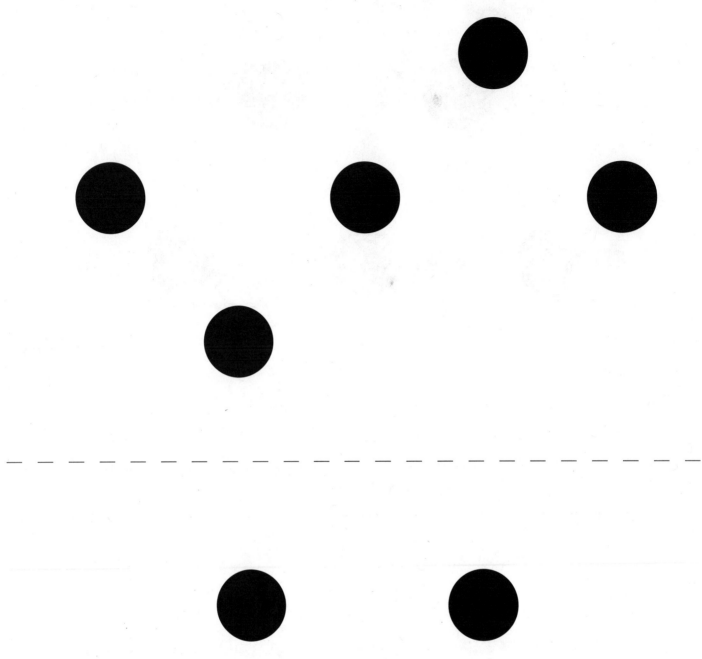

BLM 12

Dot Cards

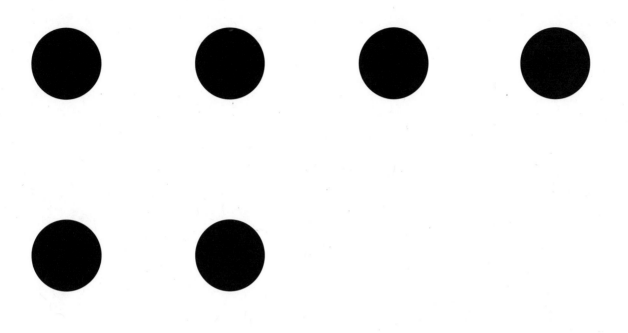

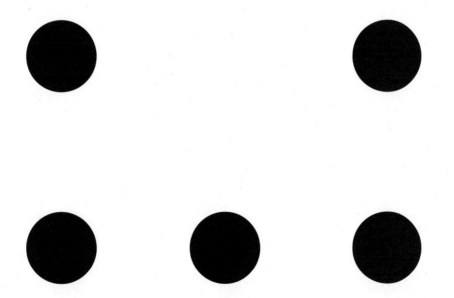

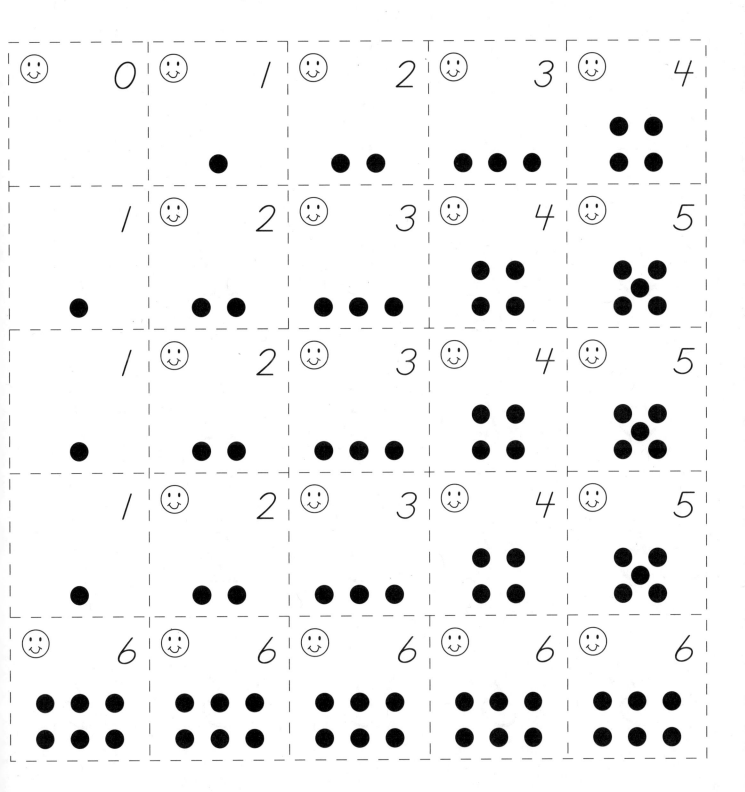

☺ 4	☺ 3	☺ 2	☺ 1	☺ 0
☺ 5	☺ 4	☺ 3	☺ 2	☺ 1
☺ 5	☺ 4	☺ 3	☺ 2	☺ 1
☺ 5	☺ 4	☺ 3	☺ 2	☺ 1
☺ 6	☺ 6	☺ 6	☺ 6	☺ 6

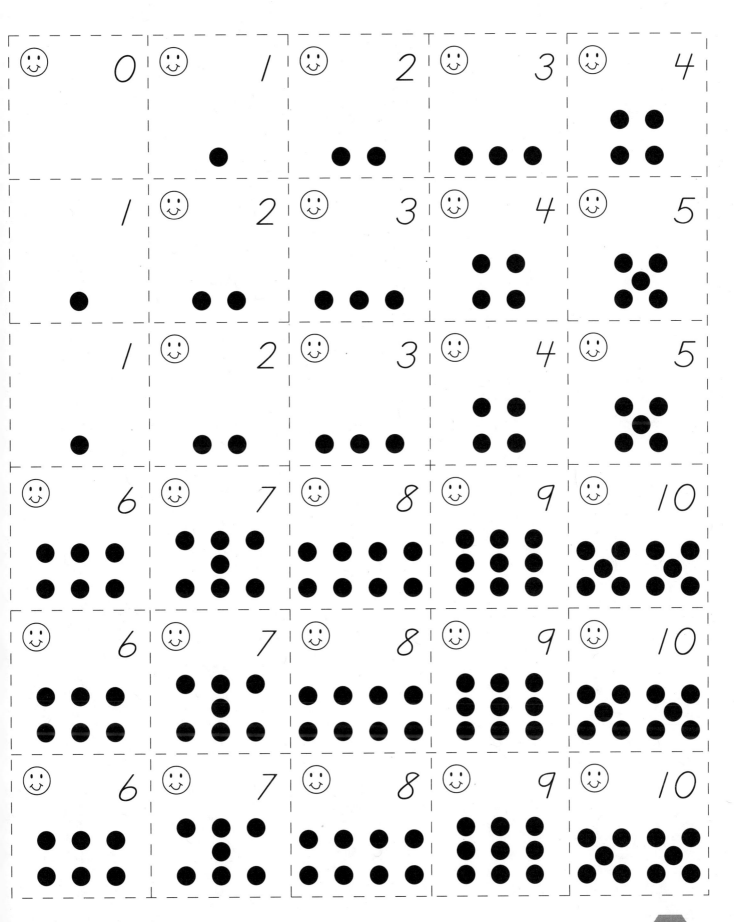

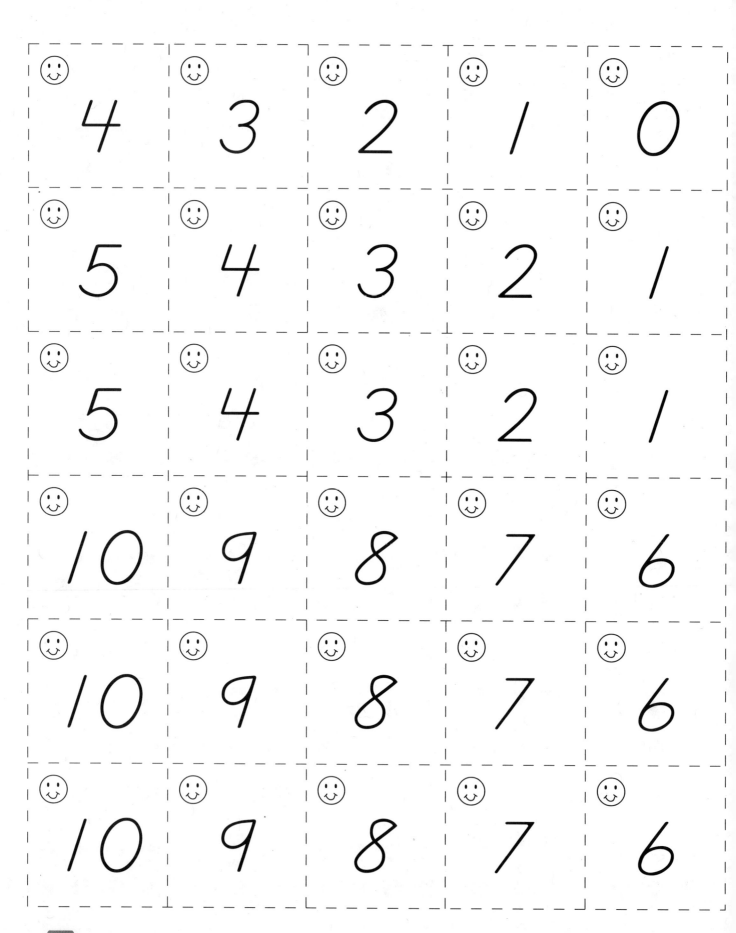

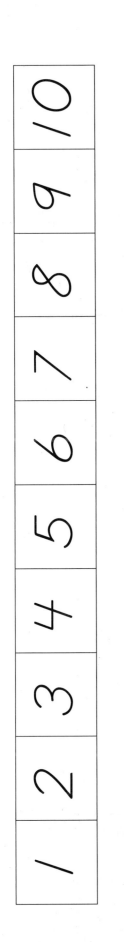

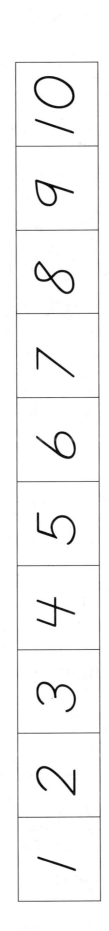

1	2	3	4	5	6	7	8	9	10
1	2	3	4	5	6	7	8	9	10
1	2	3	4	5	6	7	8	9	10

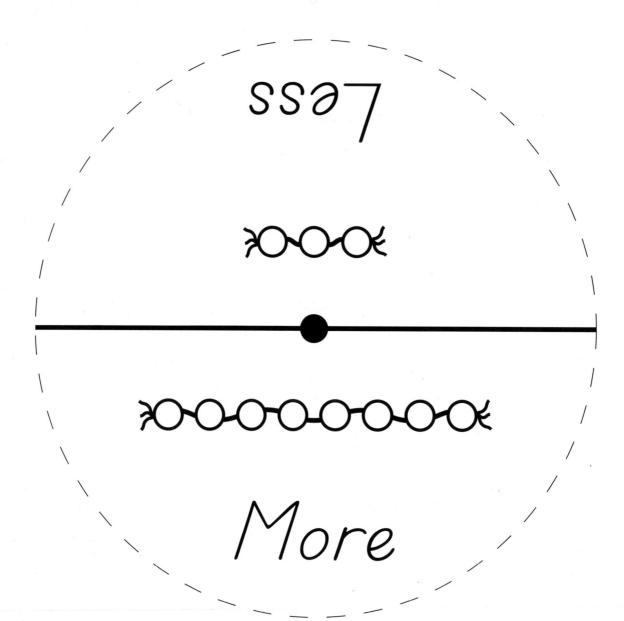

Making the More/Less Spinner:

1. Copy spinner picture on this page onto card stock.

2. Poke a hole through the center of the circle.

3. Open a paper clip and push one end through the hole.

4. To keep the spinner from coming apart and to keep the children from sticking themselves, wrap a tiny piece of masking tape around the end of the paper clip. Make sure the tape does not keep the spinner from spinning.

5. Turn the spinner over, and tape the base of the clip to the back of the spinner to secure the clip.

Less

More

Pattern Train Outlines

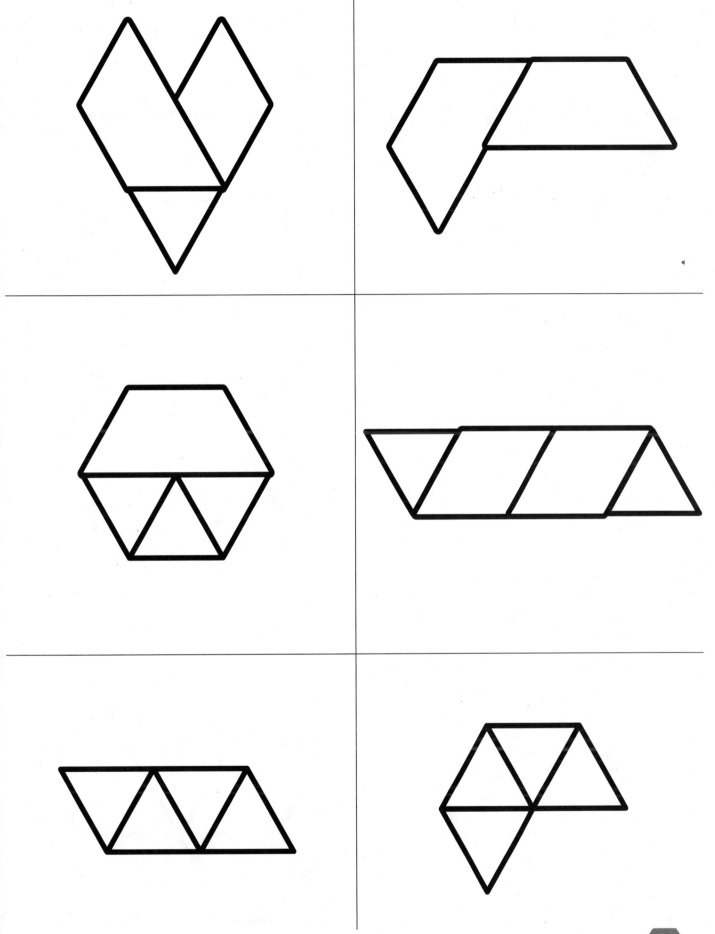

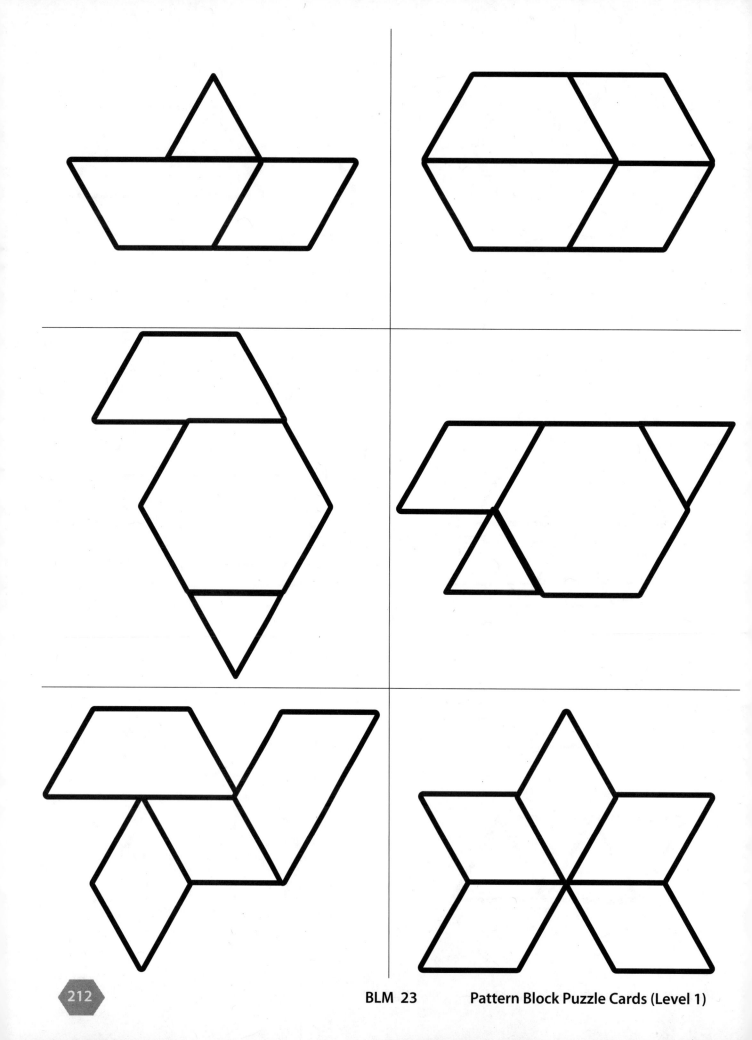

BLM 23   Pattern Block Puzzle Cards (Level 1)

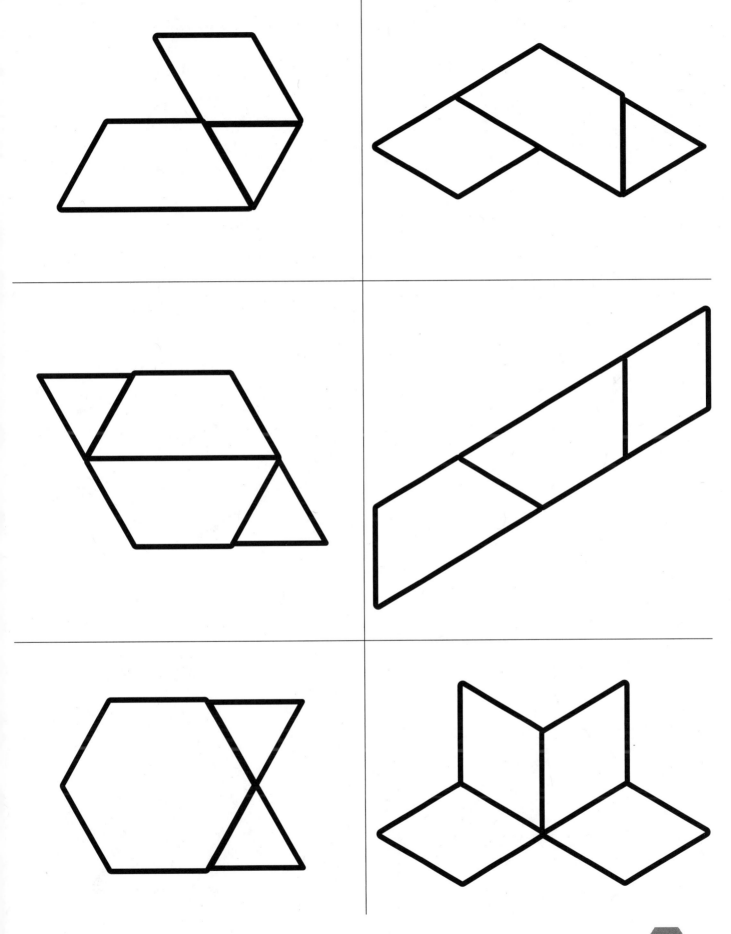

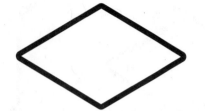

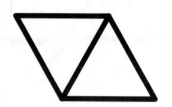

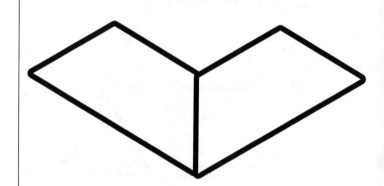

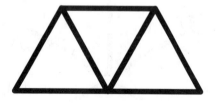

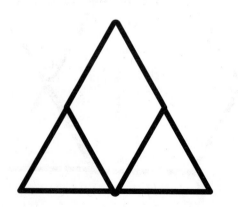

BLM 25   **Pattern Block Puzzle Cards (Level 1)**

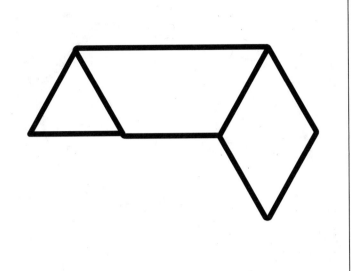

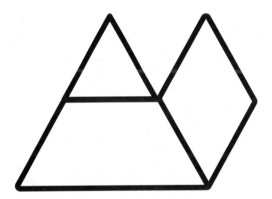

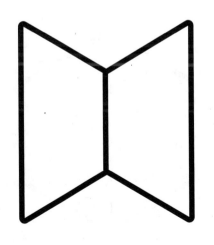

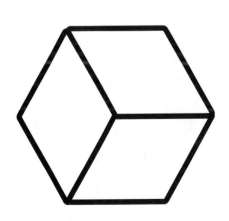

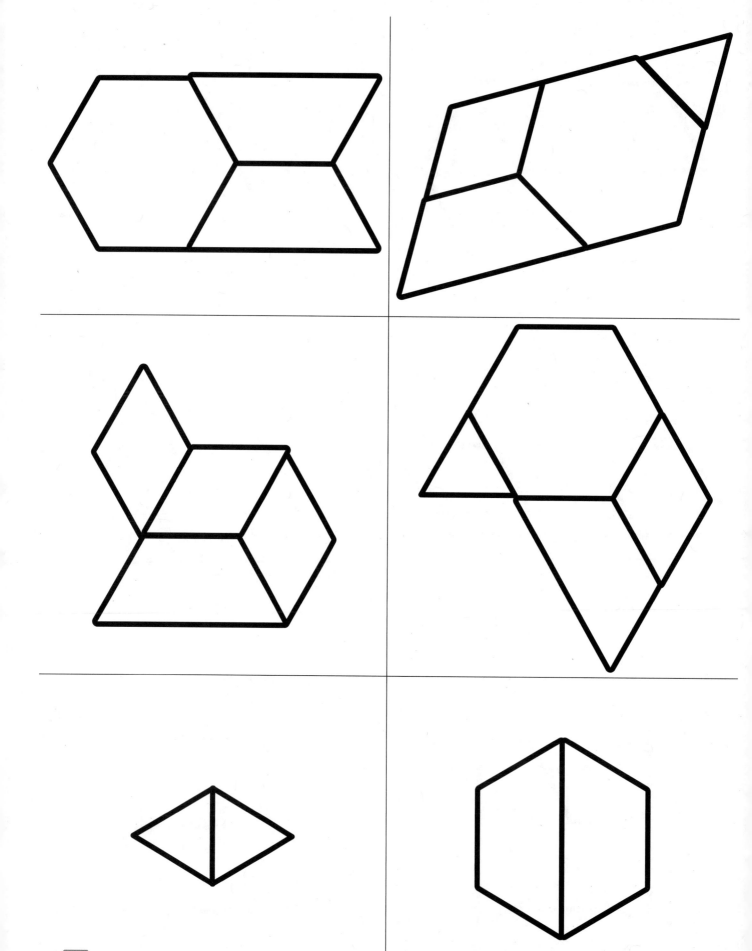

BLM 27 **Pattern Block Puzzle Cards (Level 1)**

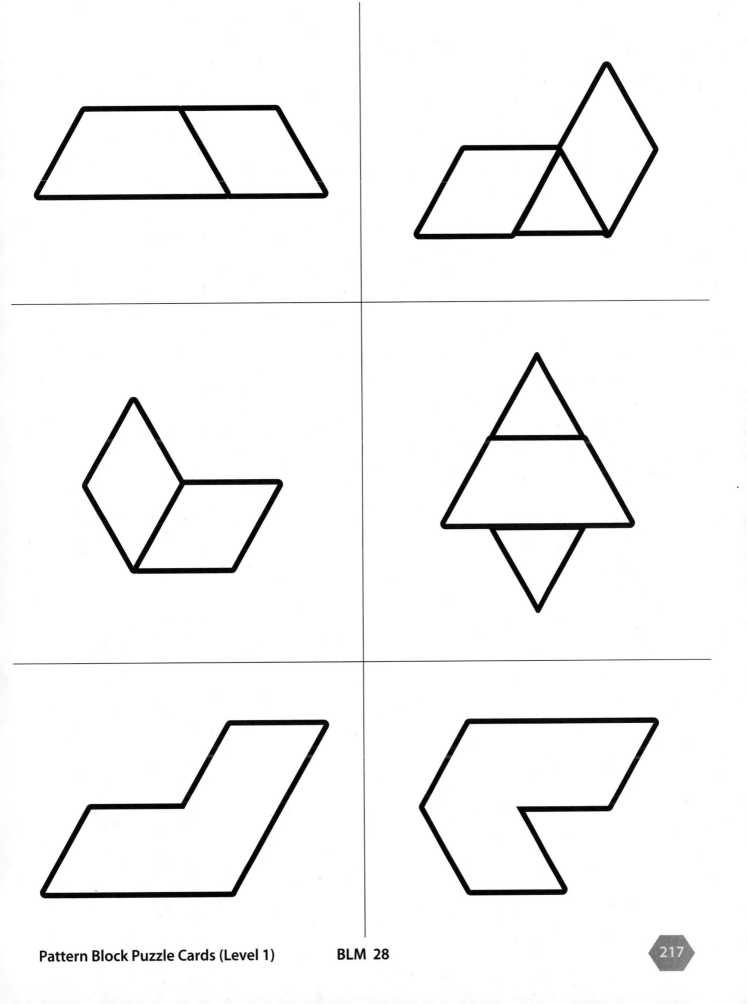

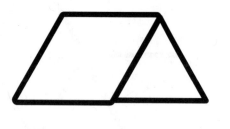

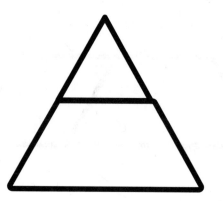

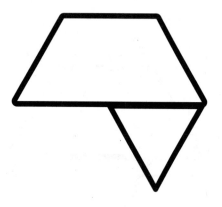

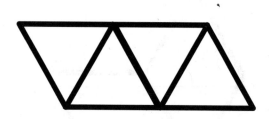

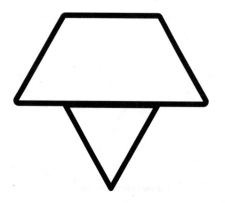

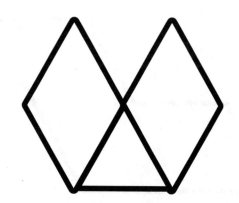

Pattern Block Puzzle Cards (Level 1)

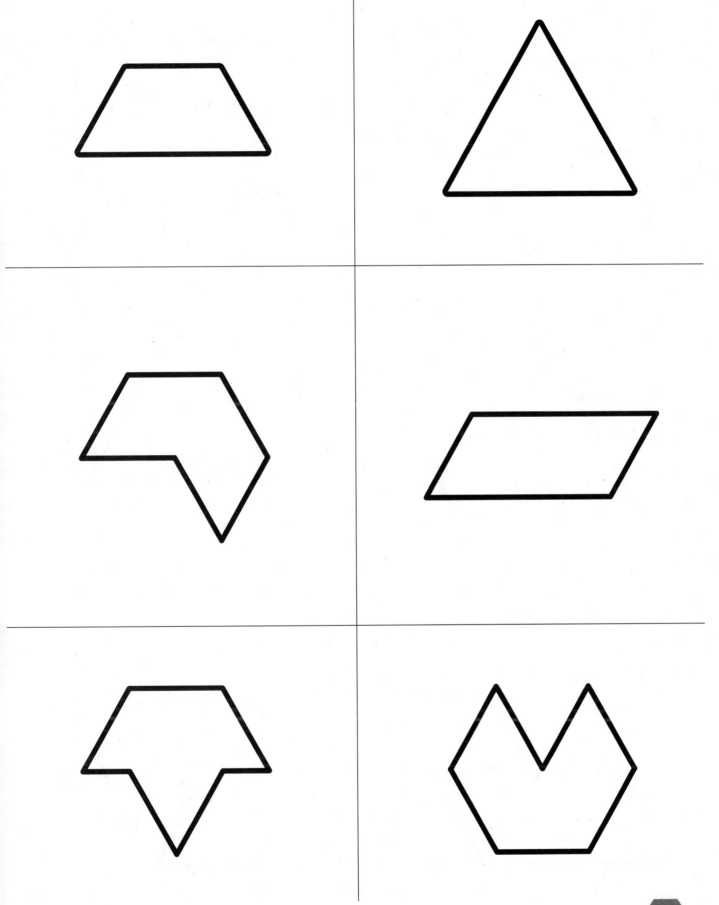

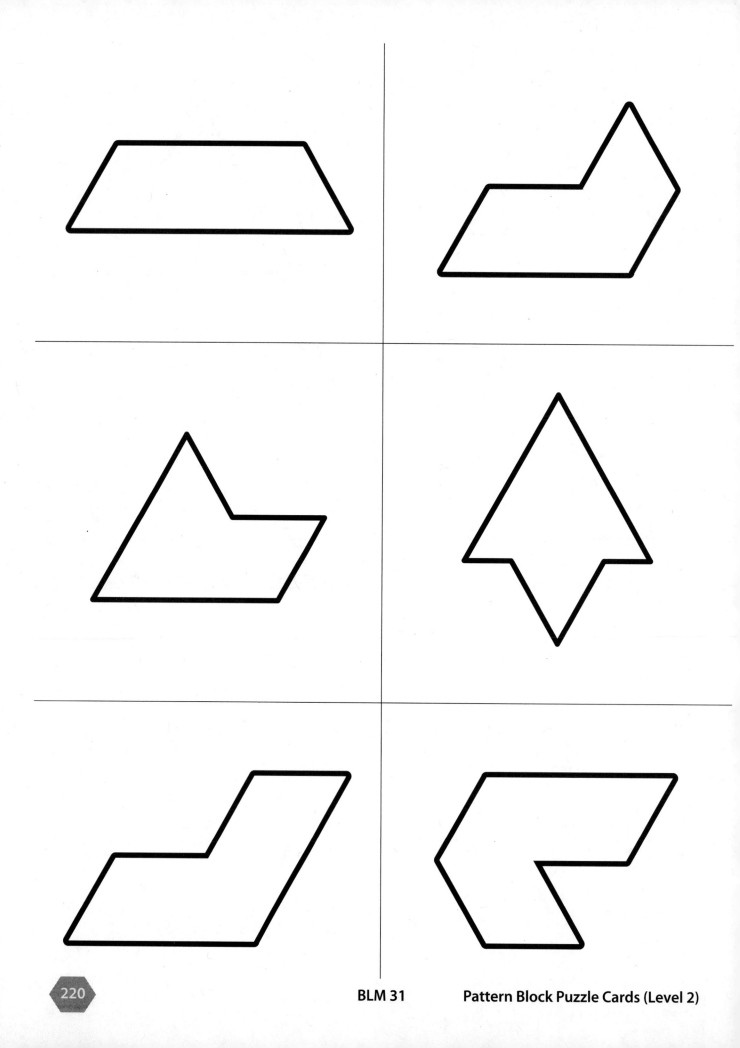

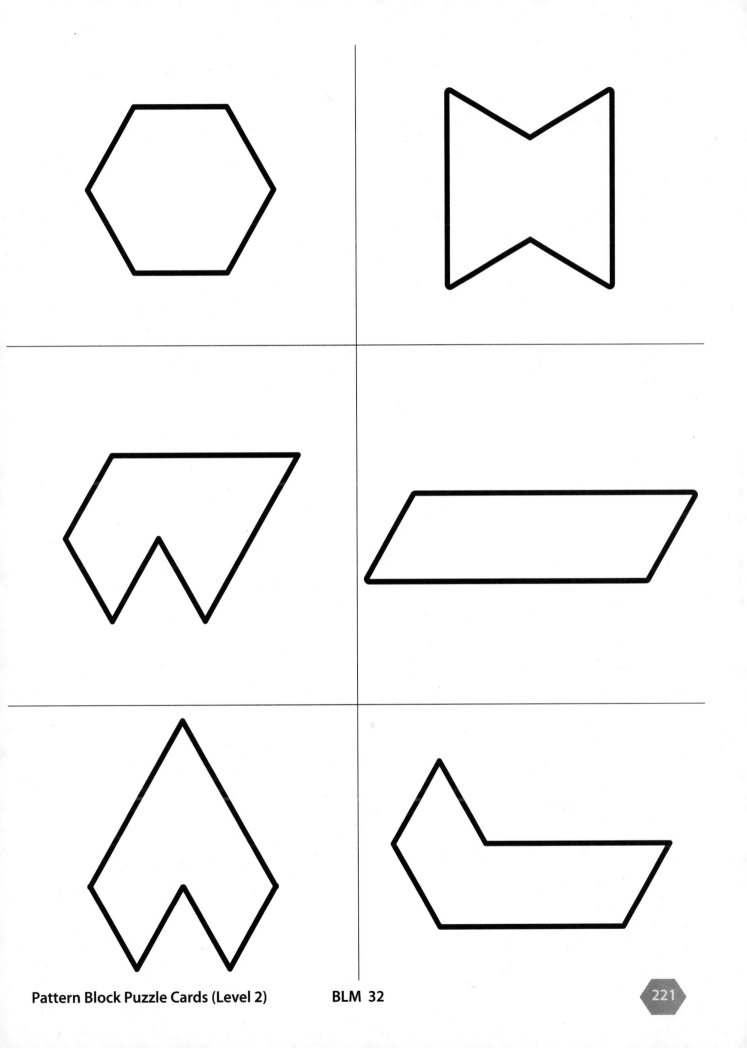

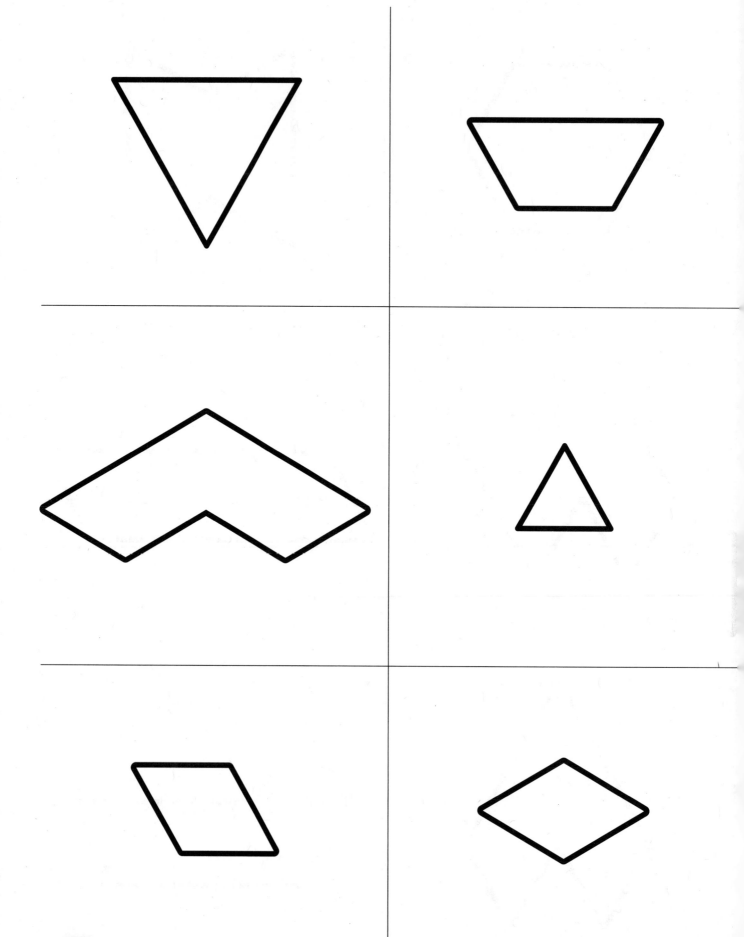

BLM 33   Pattern Block Puzzle Cards (Level 2)

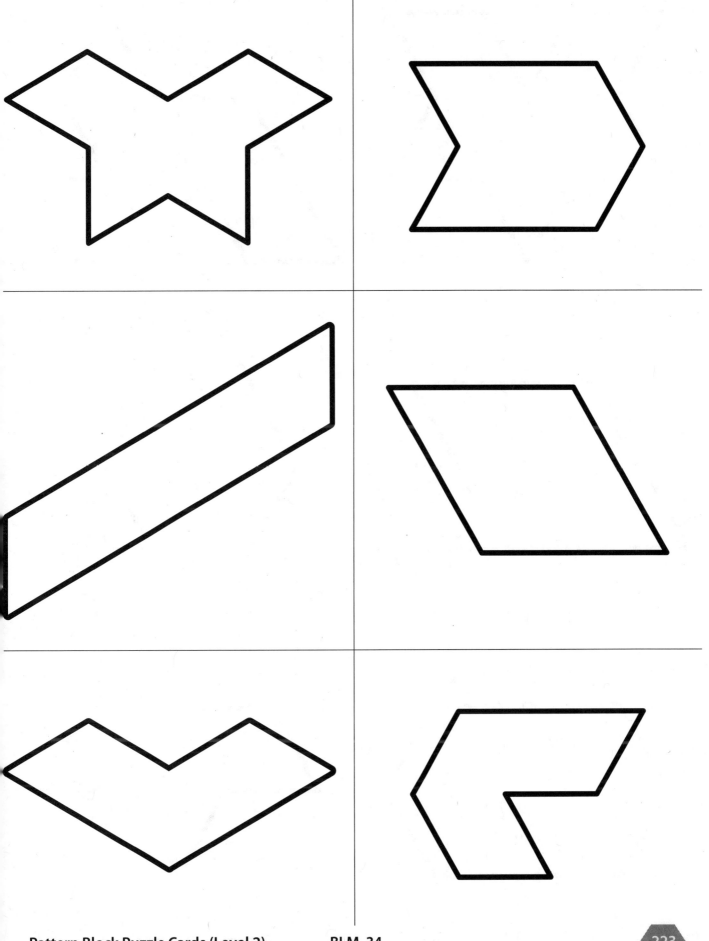

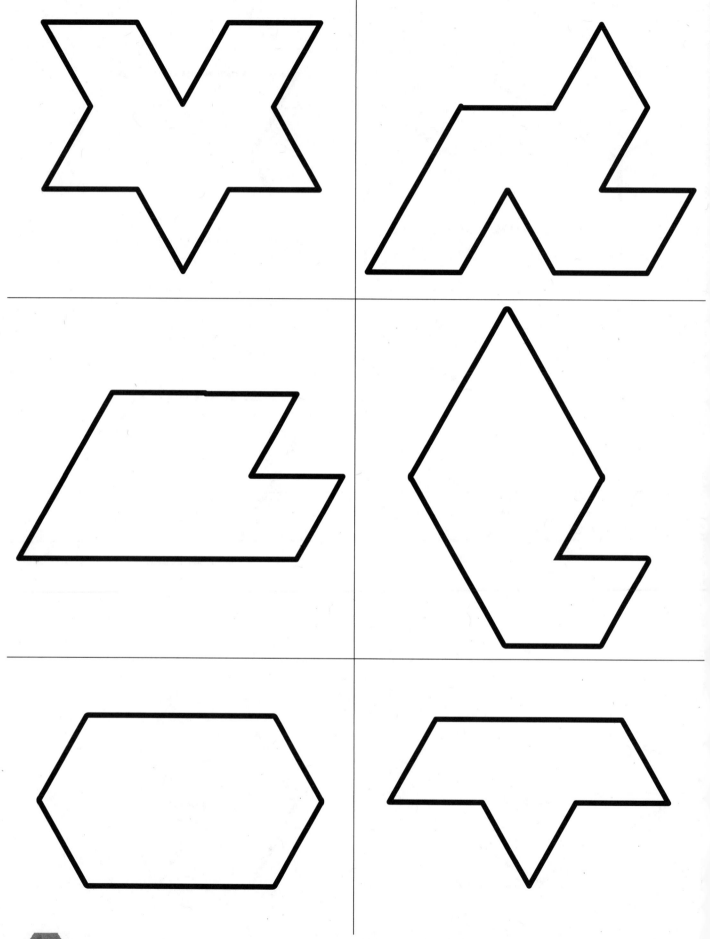

BLM 35   Pattern Block Puzzle Cards (Level 2)

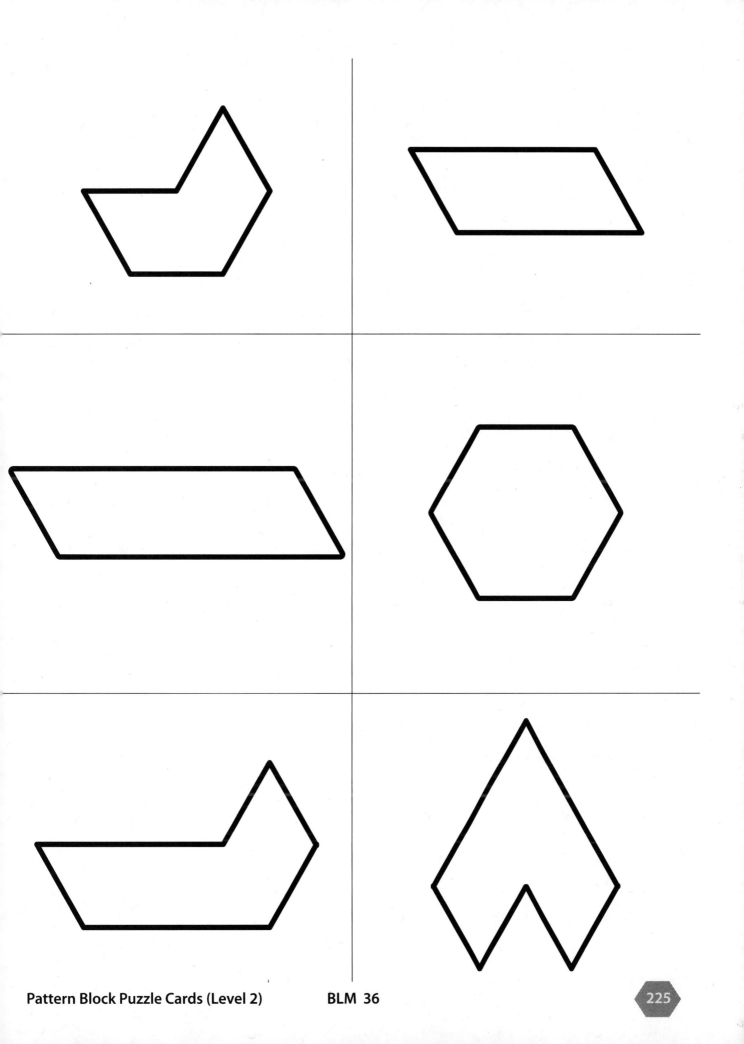

BLM 37   **Pattern Block Shape – Green**

228

Pattern Block Shape – Blue

Pattern Block Shape – Yellow   BLM 40

229

BLM 41

Pattern Block Shape – Red

ALSO BY KATHY RICHARDSON

Assessing Math Concepts Series

This K–3 formative assessment series of nine books assesses counting, comparing, addition, subtraction, and place value concepts.

- Counting Objects
- Changing Numbers
- More/Less Trains
- Number Arrangements
- Combination Trains

- Hiding Assessment
- Ten Frames
- Grouping Tens
- Two-Digit Addition and Subtraction

Developing Number Concepts Series

A complete K–3 curriculum featuring hands-on activities that help young children see and feel math! Based on the author's years of research, each book provides simple but meaningful activities to give students foundational math experiences.

- Book 1: Counting, Comparing, and Pattern
- Book 2: Addition and Subtraction
- Book 3: Place Value, Multiplication, and Division

Understanding Numbers Series

A three-book series (Place Value, Addition & Subtraction, and Understanding Decimals) consisting of task cards designed to be used in learning "stations" to develop computational fluency in grades 3–5.

Understanding Geometry

A teacher resource book to help young children develop an in-depth understanding of 2D and 3D geometry.

——————— To order, call 800-458-0024 or visit www.didax.com.———————

MATH PERSPECTIVES TEACHER DEVELOPMENT CENTER

Math Perspectives offers a wide range of professional development institutes, workshops, and seminars for teachers to continuously improve the teaching and learning of mathematics in their classrooms. For more information, call 360-715-2782 or visit www.mathperspectives.com.